1990

Chemical Structure Information Systems

ACS SYMPOSIUM SERIES **400**

Chemical Structure Information Systems

Interfaces, Communication, and Standards

Wendy A. Warr, EDITOR
ICI Pharmaceuticals

Developed from a symposium sponsored
by the Division of Chemical Information
at the 196th Meeting
of the American Chemical Society,
Los Angeles, California,
September 25–30, 1988

American Chemical Society, Washington, DC 1989

Library of Congress Cataloging-in-Publication Data

Chemical structure information systems.

(ACS Symposium Series, ISSN 0097–6156; 400).

"Developed from a symposium sponsored by the Division of Chemical Information at the 196th Meeting of the American Chemical Society, Los Angeles, California, September 25–30, 1988."

Bibliography: p.

Includes index.

1. Chemical structure—Data processing—Congresses. 2. Information storage and retrieval systems—Chemical structure —Congresses.

I. Warr, Wendy A., 1945–   . II. American Chemical Society. Division of Chemical Information. III. American Chemical Society. Meeting (196th: 1988: Los Angeles, Calif.). IV. Series

QD471.C517   1989   541.2′2′0285   89–14988
ISBN 0–8412–1664–9

ACS Symposium Series

M. Joan Comstock, *Series Editor*

1989 ACS Books Advisory Board

Foreword

The ACS SYMPOSIUM SERIES was founded in 1974 to provide a medium for publishing symposia quickly in book form. The format of the Series parallels that of the continuing ADVANCES IN CHEMISTRY SERIES except that, in order to save time, the papers are not typeset but are reproduced as they are submitted by the authors in camera-ready form. Papers are reviewed under the supervision of the Editors with the assistance of the Series Advisory Board and are selected to maintain the integrity of the symposia; however, verbatim reproductions of previously published papers are not accepted. Both reviews and reports of research are acceptable, because symposia may embrace both types of presentation.

Contents

Preface..ix

1. **Introduction**.. 1
   Wendy A. Warr

2. **Design Considerations for Multipurpose Structure Files**................ 10
   Richard S. Hong

3. **Information Integration in an Incompatible World**..................... 18
   Dennis H. Smith

4. **Chemical Structure Browsing**.. 41
   Alexander J. Lawson

5. **Integration and Standards: Use of a Host Language Interface**........... 50
   A. Peter Johnson, Katherine Burt, Anthony P. F. Cook,
   Kevin M. Higgins, Glen A. Hopkinson, and Gurmaj Singh

6. **A Standard Interface to Public, Corporate, and Personal Files**.......... 59
   John L. Macko and James V. Seals

7. **Towards the Universal Chemical Structure Interface**................... 68
   William G. Town

8. **A Universal Structure/Substructure Representation for PC-Host
   Communication**... 76
   John M. Barnard, Clemens J. Jochum, and Stephen M. Welford

9. **Chemical Structure Information at the Bench: A New Integrated
   Approach**.. 82
   Robert M. Olszewski, Everett A. Bruce, Craig Leilous, and
   Rudy Potenzone, Jr.

10. **Building a Comprehensive Chemical Information System**.............. 89
    Jean-Pierre Gay, Guillaume Auneveux, and Françoise Chabernaud

11. **The Standard Molecular Data Format (SMD Format) as an Integration Tool
    in Computer Chemistry**.. 105
    H. Bebak, C. Buse, W. T. Donner, P. Hoever, H. Jacob, H. Klaus,
    J. Pesch, J. Roemelt, P. Schilling, B. Woost, and C. Zirz

12. **The Effort To Define a Standard Molecular Description File Format**..... 118
    John S. Garavelli

INDEXES

Author Index.. 127

Affiliation Index.. 127

Subject Index... 128

vii

Preface

In the last few years many computer systems have appeared which allow graphics entry of chemical structures. At first, structure entry required an expensive graphics terminal. As cheaper graphics terminals came on the market and there were more sales of suitable software, more and more users became familiar with drawing chemical structures on a screen. Once microcomputers became widely available there was a great increase in the number of software packages on offer to research workers and information professionals. The healthy competition fostered by the advent of microcomputer packages has, on the whole, been beneficial to the end-user. However, the multiplicity of available systems does lead to one major problem: the end-user is not prepared to learn and use several different methods of drawing chemical structures. He, or she, wants to be able to use the same drawing method to prepare scientific reports and to access public, corporate and personal data bases. Not all end-users would agree on the preferred drawing method, or "front-end", and it is undesirable for any vendor to have a monopoly, but there is obviously much scope for building interfaces between the most popular drawing methods and the major sources of chemical structure information.

In the ideal world, neither commercial pressures nor technical complications should prevent a user from using his or her preferred interface to access multiple public, corporate and personal chemical structure data collections.

The "universal interface" is thus an exciting objective, but, related to it, and equally important, is the need for standards which allow the transfer of data between many different chemical structure handling systems.

This book directly addresses the problems of interfaces, communication, and standards. It is authored by vendors and users of a variety of chemical information systems, for example, for scientific document production, substructure searching, reaction indexing and molecular modeling.

My own, introductory, chapter deals with the development of standards for chemical structure representation. I have given copious literature references which should be helpful to readers of those chapters which have fewer references.

The reader who is particularly concerned with information technology, and software and hardware standards, would be well advised to read Chapter 3.

The book does not attempt to address the issue of standards for electronic publishing. However, the reader may be interested to know the electronic methodology used in producing the book. All but two of the chapters were submitted to me on floppy disks. The remaining two were downloaded from the Chemical Abstracts Service electronic mail system. Only one floppy disk could not be converted to ASCII and reprocessed. That chapter had to be word processed from scratch by my secretary. All the other chapters were sent electronically to her using the DEC All-in-1 office automation system in use at ICI. The earliest chapters unfortunately required careful checking because the terminal emulation software used did not have an error checking protocol and occasional clauses were duplicated or omitted. After this experience, we used the Kermit protocol to transmit later chapters.

My secretary edited all the chapters (according to my instructions on a hard copy) on a Wordplex machine using the WPS+ word processor. A Wordplex floppy disk of each chapter was sent for typesetting (with a hard copy indicating "problems" such as special fonts, umlauts etc.). The floppy disks were re-converted to ASCII and typeset on a Linotron 202.

Unfortunately we were not able to process "compound documents". Only the text of the book was reprocessed and typeset electronically. All diagrams and figures had to be resized from hard copy and inserted.

Finally, I sent a photocopy of the actual printed version to each author. However, I recognize that I myself must bear responsibility for any errors that may have crept in during the processing of the camera-ready "manuscript" sent to ACS.

Acknowledgments

I would like to thank all the people in the ACS Division of Chemical Information who helped to make the original symposium possible and, in particular, Joanne Witiak, the chairperson of the program committee. I am also grateful to all the authors in this book for their efforts in writing, and rewriting, the chapters within the imposed deadlines. Three people in Information Services Section at ICI Pharmaceuticals deserve special mention: Frank Loftus who translated and transmitted electronic files for me and also redrafted some diagrams; Madeline Gray who traced and checked numerous literature references; and my secretary, Mary Burgess, who has put more effort into the book than any of us. I would also like to thank Hope Services (Abingdon) Ltd., for their efficient typesetting and helpful advice.

WENDY A. WARR
April 1989

Chapter 1

Introduction

Wendy A. Warr
ICI Pharmaceuticals, Mereside, Alderley Park,
Macclesfield, Cheshire SK1O 4TG, England

Over the past twenty years there has been a trend away from user-hostile chemical information systems towards systems with user-friendly, graphics interfaces, which allow the scientist to use his preferred method of communication, that is, the chemical structure diagram. The abundance of chemical structure interfaces nowadays actually causes a new problem: there are too many systems for the user to learn and there are too few standards.

Chemical structure handling by computer in the 1960s required very specialized expertise.

During the 1970s there was a boom in Information Technology and great strides were made in chemical information systems, but such systems were usually beyond the reach of the end-user.

The advent of computer graphics (*1,2*), the proliferation of personal computers (*3*) and the development of relatively user-friendly software has brought chemical data base building, structure and substructure searching, and chemical report production within the reach of the average chemist.

The problem is no longer the lack of systems and data bases but rather the proliferation of systems which cannot be linked in a seamless manner. Drawing the same input structure into more than one software interface is a waste of time. The user will also no longer tolerate "sneakernet" (the information is put onto a tape or floppy disk and someone in sneakers runs down the corridor with it to the next machine). The systems expert has no time to write multiple conversion routines. There is quite obviously a need for seamless interfaces and good communication. Behind these lies the need for standards.

Early Attempts at Standardization and Interfacing

As far as the chemist is concerned the standard, and preferred way, of representing and communicating chemical structure information is the two-dimensional chemical structure diagram. In the early days, the hardware and software available could not handle such diagrams. If we overlook fragment codes and punch card technology (since such methods did not represent the full topology of a molecule), the earliest methods for handling chemical structures involved chemical nomenclature or line notations.

Many line notations have been suggested but only three gained significance: the

0097–6156/89/0400–0001$06.00/0
© 1989 American Chemical Society

SMILES notation (Simplified Molecular Input Line Entry System) used in the Pomona College MedChem project (4);the Dyson notation (5) adopted as a standard by the International Union for Pure and Applied Chemistry (IUPAC); and Wiswesser Line Notation (WLN), which became the actual standard used in the chemical industry (6). The rules of WLN were "standardized" by the Chemical Notation Association (CNA), but the two major WLN data bases, Index Chemicus Registry System, ICRS, from the Institute for Scientific Information and the Commercially Available Organic Chemicals Index, CAOCI (7) used somewhat different conventions.

The ICI CROSSBOW (Computerised Retrieval of Organic Structures Based on Wiswesser) system (8) had a large number of users worldwide. It used not only WLN but also fragment codes and connection tables derived from WLN. The CROSSBOW connection table was bond-implicit (9) as opposed to the bond-explicit connection table used by Chemical Abstracts Service (CAS). The CAS Registry System (10) was (and still is) based on nomenclature, registry numbers, and connection tables rather than line notations. A number of European companies bought Chemical Abstracts data for use in-house and were able to access both internal (corporate) and literature information with proprietary systems (11). ICI converted the ICRS data base into a CROSSBOW-compatible form and was thus able to use the same technology for searching a literature data base as well as ICI in-house chemical structures (12). However, the Chemical Abstracts data base was not accessible this way, despite experiments in interconversion of CAS and CROSSBOW connection tables (13).

The Emergence of Graphics Systems for Chemical Structure Searching

The earliest system which allowed substructure searching involving chemical connectivity input and structure display, was the National Institutes of Health/ Environmental Protection Agency Chemical Information System, the NIH/EPA CIS (14). Its Structure and Nomenclature Search System (SANSS) eventually allowed access to a large range of public data bases. The system allowed access from teletype terminals and was not graphics-based.

The Chemical Abstracts Registry File became substructure searchable online in the early 1980s, first by means of the DARC (Description, Acquisition, Retrieval and Correlation) system (15) as implemented by Télésystèmes and, soon afterwards, in the CAS ONLINE Service (16).

The evolution of molecular graphics (1) is described in an earlier ACS Symposium Series book (17) which acts as an interesting precursor to this present volume. Chemical reaction systems such as LHASA (Logic and Heuristics Applied to Synthetic Analysis) (18) and SECS (Simulation and Evaluation of Chemical Synthesis) (19) had long used graphics but it was some time before the first in-house, proprietary system appeared, attracting much interest in the chemical and pharmaceutical industries. This was Upjohn's Compound Information System, COUSIN (20–21).

Neither COUSIN nor the CAS ONLINE Messenger software were portable or commercially available. From the early 1980s there was a big demand for user-friendly, interactive access to in-house, chemical structure data bases. The market leaders became MACCS (22–23) marketed by Molecular Design Ltd (MDL) and

DARC in-house (*15,23*) marketed by Télésystèmes. OSAC (Organic Structures Accessed by Computer) from the ex-Leeds University team ORAC Ltd, appeared rather later (*24*).

Many organizations acquiring these new systems already had files of structures on chemical typewriters or encoded in WLN and wanted to generate MACCS data bases or DARC connection tables automatically. Interconversion thus became the rage of the mid-1980s. The Chemical Structure Association's first publication was the proceedings of the CNA(UK) seminar on interconversion held at the University of Loughborough in March 1982. Elder's DARING program to convert WLNs to connection tables is described therein. This program has been widely used but further software is required to convert the DARING connection tables to MACCS, or other, versions, and to generate structure coordinates needed for the actual graphics display of the structures (*25*).

Some companies have written proprietary algorithms to allow structures drawn in MACCS to be entered to the Pomona College MedChem system for structure activity relationships.

Integration of Chemical Structure Data with Property Data

Once many in-house, chemical structure data bases had been built, the users began to realize that it was more efficient to use commercially-available structure handling software for structures alone (or structures and a minimal amount of related property data) and to take advantage of data base management systems to handle property data. Molecular Design Limited (MDL) and Télésystèmes modified their MACCS and DARC software, respectively, to allow for the appropriate interfacing of structures and data.

Chapters 5 and 10 in this book discuss such interfaces (and related topics).

The era of integration of structures with text (*17*) was hardly beginning.

Proliferation of Incompatible Systems in the 1980s

By the mid-1980s the advent of the microcomputer had started to make a big impact on the world of chemical information (*3*). Chemical structure drawing packages such as ChemDraw (*26–27*); the Wisconsin Interactive Molecular Processor, WIMP (*26*); and Molecular Presentation Graphics, MPG (*26*) could be used for document production. Multipurpose, connection table based programs such as PSIDOM (Hampden Data Services' Professional Structure Image Database on Microcomputers) (*26,28*) and CPSS (Molecular Design Limited's Chemists' Personal Software Series) (*26,28–29*) were available for document production, data base building and substructure searching. The structure drawing technology in some of these microcomputer-based packages was much more sophisticated and user-friendly than in previous graphics interfaces such as that to CAS ONLINE. The advantages of using a microcomputer package as a "front-end" to the online systems CAS ONLINE, DARC and CIS were discussed (*3*).

The first such front-end to appear was Fein-Marquart's SuperStructure (*30*), which allowed graphics chemical structure input for accessing the CIS data bases.

Meanwhile, a multiplicity of chemical and pharmaceutical companies involved in the Molecular Design Limited Software Users Group as users of MACCS and

CPSS started to demand facilities for downloading structures from CAS ONLINE into MACCS in-house data bases. In an unpublished survey carried out in 1987, the author of this chapter established that members of the Molecular Design Limited Software Users Group were even more interested in uploading i.e., drawing a structure query using MACCS methodology and sending it up the line to search the CAS ONLINE data base. Interestingly the users had a preference for MACCS over CPSS for this process. Their views on the user-friendliness of structure input packages were heavily colored by their *familiarity* with MACCS and CAS ONLINE. Indeed, there was an abysmal ignorance of the wealth of structure drawing facilities available both in the United States and in Europe. It is hoped that a recent ACS Professional Reference Book (*26*) will rectify this situation. In that book about 70 different software packages for personal computers are described.

In this author's company, by the end of 1987, information scientists were faced with mastering a large range of methods for drawing chemical structures both on graphics terminals and on personal computers. Examples were:

1. MACCS (for in-house chemical data bases).
2. CAS ONLINE and DARC (for searching the chemical literature).
3. ORAC (Organic Reactions Accessed by Computer, the reaction indexing software written at Leeds University and marketed by ORAC Ltd.) (*31*).
4. PsiGen (the structure drawing module of Hampden Data Services' PSIDOM software) (*26*).
5. PsiORAC (the PsiGen interface to ORAC).
6. ChemDraw (the Macintosh software package from Cambridge Scientific Computing which is the ICI chemist's preferred method for drawing high-quality structures for pasting into documents) (*26–27*).
7. SANDRA (the Structure and Reference Analyzer marketed by Springer Verlag for help with use of the Beilstein Handbook) (*26,32*), and
8. TOPFRAG (the Topological Fragment Code Generator program for chemical structure access to the Derwent patents data base) (*26*).

Other companies might cite different examples but their listings would be equally long.

It is obvious that the end-user will not be prepared to learn this multiplicity of ways of inputting a chemical structure.

In the ideal world, neither commercial pressures, nor technical complications, should prevent a user from drawing a chemical structure the way he or she wants to and accessing any personal, corporate or public file with that structure (or substructure). In practice, both commercial and technical factors cause severe limitations. This is the problem at which this book is aimed.

Front-End Software

Since the book was conceived new front-ends such as STN Express (*26,33*), MOLKICK (*26,34*) and DARC CHEMLINK (*26*) have appeared on the market. All are IBM PC based graphics packages which allow offline query formulation followed by connection to a host computer and uploading of the query for online searching. Offline query formulation means that an end-user can avoid the so-called "taxi-meter syndrome" that is a consequence of incurring online charges if formulating a search strategy while connected to the host computer.

DARC CHEMLINK (from Télésystèmes) is a program which allows offline formulation of queries for submission to DARC data bases. Its structure drawing interface is intentionally very similar to that of the DARC system online.

MOLKICK (sold by Springer Verlag), which is described in Chapter 8 of this book, is a memory-resident query editor which converts a structure to a string suitable for uploading to Beilstein/Softron, CAS or DARC data bases. Its chemical structure input is very similar to Beilstein's MOLMOUSE software (*35*).

STN Express (the Chemical Abstracts Service approved front-end to CAS ONLINE) is touched upon in Chapters 6 and 7. It, also, allows the user to formulate queries offline and provides help with special difficulties such as tautomerism. In addition it has a Guided Search module for novices in the STN Command Language and Boolean logic. Structure drawing is compatible with PsiGen.

MOLKICK allows the searcher to use his own preferred communication software. STN Express incorporates its own communications package, but in so removing the user's freedom of choice, it does impose an error-checking protocol which has considerable advantages.

With in-house data bases, the financial advantages of offline query formulation are less obvious, but there is still a business need for uploading structures from a PC to a corporate data base or downloading from the mainframe to the PC. MDL's CPSS package supplies such a front-end to data bases under MDL's MACCS software.

Télésystèmes have devised a way of capturing DARC structure vectors (not connection tables) for display with the popular ChemDraw package on a Macintosh, but (as yet) ChemDraw substructure searchable data bases are not a possibility.

Unfortunately downloading is not yet possible with STN Express. There is a package called CASKit (*26*), unsupported by both MDL and CAS, which captures the vectors for CAS ONLINE output structures, converts them into a graphics metafile and then converts the metafiles into MACCS-compatible connection tables plus structure coordinates.

The writer, or user, of a PC package which is an unsupported interface to software from a major vendor, faces obvious dangers. The vendor of the host system may, maliciously or unknowingly, change minor features of his file structure or command language, rendering the PC interface inoperable.

Commercial Considerations

The expression "major vendor" was used advisedly in the above. In most of the collaborative ventures seen so far a major vendor uses the products of a small software house or develops software himself. From the viewpoint of the users, it is unfortunate that major vendors cannot cooperate with each other. The prospects for a supported MACCS interface to the CA Registry File are still not good.

Of the PC-based packages, PSIDOM is one that has been aimed especially at collaborative software developments. There are PsiGen interfaces to the CA Registry File (PsiCAS in STN Express) to ORAC (PsiORAC) to Derwent data bases (in the programs TOPFRAG and TORC which convert structures to Derwent fragment codes) and to DARC in-house data bases. We will leave the reader to speculate whether it is really technical difficulties that prevent the appearance of "PsiMACCS".

It is commercial considerations that have led to the abundance of IBM PC based software for chemical structure handling and the limited number of packages for use on the Apple Macintosh. However, the user base for the Macintosh is increasing and this situation will change.

Standardization in the Macintosh environment is such that the user of programs such as ChemDraw could easily adapt to other well-designed Macintosh structure drawing packages. The appearance of "ChemDraw-like" front-ends in the near future is a certainty. (There is already a Macintosh-based front-end to the CA Registry File, called ChemConnection, marketed by Softshell of Henrietta, New York, but not supported by CAS.)

Technical Considerations

Computer graphics standards are discussed in an earlier ACS Symposium Series Book (1,2). In the present book, the reader is particularly referred to Chapter 3, by Smith, for a detailed exposition of trends and standards in hardware, operating systems and environments, and applications software. Smith also examines the implications for chemical information.

Standard File Structures

One could state with wry humour that the good thing about standards is the number of them that there are. This is as true in the chemical structure representation field as in other fields.

Some "standard" ways of storing and transferring chemical structures are proprietary (e.g., MDL's Molfile); others such as the JCAMP-CS format, published by the Joint Committee on Atomic and Molecular Physics, are in the public domain. Barnard (36) refers to some of them in a paper that deals with recent developments in improving the Standard Molecular Data (SMD) file format and work towards establishing it as the one standard for transfer of chemical structure information between systems. In Chapter 11 of this book, Donner et al. describe the SMD format in more detail. Garavelli, in Chapter 12, also discusses SMD, but concentrates on existing "standards" for molecular modeling systems.

Summary of Chapters

In Chapter 2 Richard Hong discusses the wide variety of uses of chemical structure information and the need for flexibility in a file format for free exchange of chemical data.

The next chapter by Dennis Smith of MDL describes hardware and software standards in detail. The reader who is particularly concerned with information technology would be well advised to read this. The chapter is a technical and technocommercial one. It is not intended simply to explain MDL's commercial position, any more than Chapter 6, by Chemical Abstracts Service authors, is supposed to state simply a CAS position. However, these contributions (and others) were invited partly because of the significance of the vendors involved. The chapter by Chemical Abstracts authors is followed by Bill Town's contribution since his company, Hampden Data Services, collaborated in the STN Express front-end to CAS ONLINE.

Sandy Lawson's contribution on chemical structure browsing (Chapter 4) may seem peripheral to the main topic of the symposium, but the algorithm he employs could, he hopes, be used for data bases other than Beilstein in future.

Other developments at the Beilstein Institute, and in particular the so-called ROSDAL string for transferring chemical structure information from PC to host computer, are described in Chapter 8.

Chapter 5 deals with integration and standards both for a reaction indexing systems (ORAC) and for structure management software (OSAC).

Some useful interfaces to the DARC system are described in Chapter 10.

Chapter 9 is concerned with Polygen's CENTRUM system as an integration tool for the various components of scientific documents.

As described earlier, Chapters 11 and 12 concern standard molecular description files.

Conclusion

The present state of the art in interfacing, communication and standards in the field of chemical structure information is rather confused. Many problems and issues are being aired but there are few answers, let alone standards. Software, some of it very useful, is nevertheless appearing and the vendors cannot afford to wait for standards to be laid down. Standards committees are notoriously slow in their deliberations. It remains to be seen whether the SMD movement will establish a standard in a reasonable time or whether a *de facto* standard will become established before then.

Literature Cited

1. Wipke, W.T. In *Graphics for Chemical Structures: Integration with Text and Data*; Warr, W.A., Ed.; ACS Symposium Series 341; American Chemical Society: Washington D.C., 1987; pp 1–7.
2. Sanderson, J.M.; Dayton, D.L. In *Graphics for Chemical Structures: Integration with Text and Data*; Warr, W.A., Ed.; ACS Symposium Series 341; American Chemical Society: Washington D.C., 1987; pp 128–142.
3. Town, W.G. In *Chemical Structures: The International Language of Chemistry*; Warr, W.A., Ed.; Springer Verlag: Heidelberg, 1988; pp 243–249.
4. Weininger, D. *J. Chem. Inf. Comput. Sci.* **1988**, *28*, 31–36.
5. Dyson, G.M. In *Chemical Information Systems*; Ash, J.E.; Hyde, E., Eds.; Ellis Horwood: Chichester, 1974; pp 130–155.
6. Baker, P.A.; Palmer, G.; Nichols, P.W.L. In *Chemical Information Systems*; Ash, J.E.; Hyde, E., Eds.; Ellis Horwood: Chichester, 1974; pp 97–129.
7. Walker, S.B. *J. Chem. Inf. Comput. Sci.* **1983**, *23*, 3–5.
8. Eakin, D.R. In *Chemical Information Systems*; Ash, J.E.; Hyde, E., Eds.; Ellis Horwood: Chichester, 1974; pp 227–242.

9. Ash, J.E. In *Chemical Information Systems*; Ash, J.E.; Hyde, E., Eds.; Ellis Horwood: Chichester, 1974; pp 156–176.

10. Dittmar, P.G., Stobaugh, R.E.; Watson, C.E. *J. Chem. Inf. Comput. Sci.* **1976**, *16*, 111–121.

11. Graf, W.; Kaindl, H.K.; Kniess, H.; Schmidt, B.; Warszawski, R. *J. Chem. Inf. Comput. Sci.* **1977**, *19*, 51–55.

12. Warr, W.A. In *Proc. CNA(UK) Seminar on Integrated Data Bases for Chemical Systems*; Chemical Notation Association (UK), 1979; pp 103–113.

13. Campey, L.H.; Hyde, E.; Haygarth Jackson, A.R. *Chem. Br.* **1970**, *6*, 427–430.

14. Heller, S.R.; Milne, G.W.A.; Feldmann, R.J. *Science* **1977**, *195*, 253–259.

15. Attias, R. *J. Chem. Inf. Comput. Sci.* **1983**, *23*, 102–108.

16. Farmer, N.A.; O'Hara, M.P. *Database* **1980**, *3*, 10–25.

17. *Graphics for Chemical Structures: Integration with Text and Data*; Warr, W.A., Ed.; ACS Symposium Series 341; American Chemical Society: Washington D.C., 1987.

18. Corey, E.J.; Wipke, W.T. *Science* **1969**, *166*, 178–192.

19. Wipke, W.T. In *Computer Representation and Manipulation of Chemical Information*; Wipke, W.T.; Heller, S.R.; Feldmann, R.J.; Hyde, E., Eds.; John Wiley and Sons: New York, 1974; pp 147–174.

20. Howe, W.J.; Hagadone, T.R. *J. Chem. Inf. Comput. Sci.* **1982**, *22*, 8–15.

21. Howe, W.J.; Hagadone, T.R. *J. Chem. Inf. Comput. Sci.* **1982**, *22*, 182–186.

22. Anderson, S. *J. Mol. Graphics* **1984**, *2*, 83–90.

23. *Communication, Storage and Retrieval of Chemical Information*; Ash, J.E.; Chubb, P.A.; Ward, S.E.; Welford, S.M.; Willett, P., Eds.; Ellis Horwood: Chichester, 1985; Chapter 7.

24. Magrill, D.S. In *Chemical Structures: The International Language of Chemistry*; Warr, W.A., Ed.; Springer Verlag: Heidelberg, 1988; pp 53–62.

25. Warr, W.A. *J. Mol. Graphics* **1986**, *4*, 165–169.

26. *Chemical Structure Software for Personal Computers*; Meyer, D.E.; Warr, W.A.; Love, R.A., Eds.; ACS Professional Reference Book; American Chemical Society: Washington, D.C., 1988.

27. Johns, T.M. In *Graphics for Chemical Structures: Integration with Text and Data*; Warr, W.A., Ed.; ACS Symposium Series 341; American Chemical Society: Washington, D.C., 1987; pp 18–28.

28. Meyer, D.E. In *Chemical Structures: The International Language of Chemistry*; Warr, W.A., Ed.; Springer Verlag: Heidelberg, 1988, pp 251–259.

29. del Rey, D. In *Graphics for Chemical Structures: Integration with Text and Data*; Warr, W.A., Ed.; ACS Symposium Series 341; American Chemical Society: Washington, D.C., 1987; pp 48–61.

30. McDaniel, J.R.; Fein, A.E. In *Graphics for Chemical Structures: Integration with Text and Data*; Warr, W.A., Ed.; ACS Symposium Series 341; American Chemical Society: Washington, D.C., 1987; pp 62–79.

31. Johnson, A.P.; Cook, A.P. In *Modern Approaches to Chemical Reaction Searching*; Willett, P., Ed.; Gower: Aldershot, 1986; pp 184–193.

32. Lawson, A.J. In *Graphics for Chemical Structures: Integration with Text and Data*; Warr, W.A., Ed.; ACS Symposium Series 341; American Chemical Society: Washington, D.C., 1987; pp 80–87.

33. Wolman, Y. *J. Chem. Inf. Comput. Sci.* **1989**, *29*, 42–43.
34. Bucher, R. *Proc. 12th Internat. Online Inf. Mtg.*, 1988, pp 183–187.
35. Jochum, C.; Ditschke, C.; Lentz, J.-P. In *Graphics for Chemical Structures: Integration with Text and Data*; Warr, W.A., Ed.; ACS Symposium Series 341; American Chemical Society: Washington, D.C., 1987; pp 88–101.
36. Barnard, J.M. *Proc. 12th Internat. Online Inf. Mtg.*, 1988, pp 605–609.

RECEIVED May 2, 1989

Chapter 2

Design Considerations for Multipurpose Structure Files

Richard S. Hong

Hawk Scientific Systems, 170 Kinnelon Road, Kinnelon, NJ 07405

Chemical structure information is used in a wide variety of different software applications. Any file format for describing a molecule or reaction must accommodate diverse uses. Many existing applications require the storage of data elements which are not addressed in current connection table formats. The adoption of an inflexible file format as a *de facto* or *de jure* standard may hamper future applications which require the free exchange of chemical data. Specific needs of existing types of applications are addressed, as are methods of accommodating future needs which are not yet known.

The computer has been a tool used by chemists for some time. However, only in the past few years has the computer evolved beyond being a specialized instrument used only by a chosen few. As in every other business, computer software used by chemists has spread into laboratories and offices far ahead of the development of any standards for the storage and exchange of data. The lack of standards is often lamented by end-users who become trapped by their inability to exchange their chemical data freely. The current chaotic situation results in end-users who use one software package for modeling, one for data base management, one for online searching, and yet another to write a report.

This discussion is taking place because this situation is a problem for many, and I hope to define the bounds of the potential solution from my personal perspective. I can only hope to raise relevant questions; finding the answers is up to you.

The development of a standard will be guided by:

1. The needs of the end-users.
2. The ability of vendors to deliver.
3. The willingness of vendors to deliver.

The development of a standard format for the exchange of chemical structure data will be the result of both cooperation and conflict between vendors and users. The users will present their needs, and the vendors attempts to meet those needs will be constrained by both technological and financial constraints.

0097–6156/89/0400–0010$06.00/0
© 1989 American Chemical Society

Defining the End-User

Sometimes it is helpful to approach a problem at the level of a 5-year-old child. Young children specialize in the one-word interrogative, "Why?" If you went home to your 5-year-old and said "Son, Mommy (or Daddy) is working on a standard format for exchanging chemical structure data", little Johnny may well say "Why?". Johnny is not as naive as those of us with years of schooling; for it is a very valid and perceptive question.

 Why are we on this quest? Presumably, someone or something will benefit from the development of a standard. However, saying that someone or something will benefit is considerably different from saying that everyone (or even most people) will benefit. So before we can proceed, we need to define the end-user community, in order that we may see just who it is that wants this "advance".

The end-user may be any of the following:

1. Mankind (through the advancement of science).
2. Institutions (companies, universities).
3. Individuals (us).

At the grandest level we can think of the end-user as mankind as a whole. In such a case, the goal of developing a standard would be to facilitate the advancement of science, with an emphasis on pushing the frontiers of chemistry. In this view, the priority applications might be advanced computational programs, with an emphasis on the ease of manipulating the data. Costs, within generous limits of reason, are secondary considerations.

We could also consider the end-user to be the institution to which we belong, whether it is a corporation, university, or a small consulting firm. Institutions tend to focus on the potential efficiencies of expediting cooperation between individuals, eliminating duplicated efforts and maintaining a collective memory. From an institutional perspective, a chemical structure standard may well be more of an administrative than a scientific problem.

It is important to realize, however, that from a corporate perspective, efficiency does not necessarily expedite science. It is possible for a corporation to consider a system which *slows* research to be more efficient if it slows down the *expense* of research at an even greater rate.

There is a tendency for professionals to be arrogant toward others in setting organizational priorities. To be sure, one hour of a chemist's time is worth more than one hour of a secretary's time. Is it worth more than four hours of a secretary or two hours of a salesperson? It is human nature for a chemist at a pharmaceutical firm, like a doctor in a hospital, or an actor in a movie, to be reluctant (on the basis of professional "rank") to make sacrifices for the good of the less powerful, even if the organization benefits as a whole.

A final view, and one to which we are prone, is to consider the individual chemist to be the end-user. In this view, we are most concerned with ourselves and our work, and therefore personal productivity applications, of whatever sort, take priority.

So in my view we have three constituencies in the development of a standard, and they are not necessarily in harmony. We must be careful to weigh the needs, abilities, and biases of each in guiding our developmental efforts.

Existing Applications Categories

To assist us in evaluating our potential "end-users", we should look at applications which exist already. It is useful to examine the nature of existing applications because they provide us with an insight into the current needs of various end-users, and the ways in which vendors meet (or fail to meet) those needs. In terms of specific applications, here is a brief (partial) list of distinct product categories which use chemical structures:

1. Computational (including 3-D modeling, etc.).
2. Data bases (single compound or reactions; public or private).
3. Publications (journals, patents, and in-house reports).

For each of these very general categories, what information needs to be maintained? The computational software usually requires the barest amount of information to start with. In most existing formats, this is simply the barebones generic connection table, consisting of one compound, exactly defined, per file.

In the category of data bases, there is a range of complexity to the data which must be maintained. In the simplest case, the single-compound-per-record data base, little information beyond the basic structure is needed. Since this minimal amount of data is sufficient for a number of applications, there will be a significant lobby in favor of limiting the format to this data. Data base software places a great premium on the size of each file. Forcing data base applications to maintain additional data will have a negative effect on their performance. Excessive verbiage places great strain on any data base.

There are two more types of file formats which add a level of complexity. In reaction data bases, the obvious differences are that multiple compounds must be stored along with their relationships (e.g., reactant, product, catalyst, etc.). Another level of complexity is added if a single diagram is used to represent an entire group of compounds. In some patent applications you see diagrams which encompass so many compounds that I would be surprised if there are not some pharmaceutical analogues which have accidentally been patented twice.

Now to this point, the data discussed has been limited to items which describe the actual physical nature of the molecule (the substance of the substance, if you will). A final, broad category of applications is one which I called "Publications", and it is the most difficult to handle. This category brings up the problem of storing the pictorial representation of the structure in addition to the structure itself. This is no small addition. While a given compound has one and only one chemical composition, it may have myriad pictorial representations.

This is the classic "style vs. substance" problem. The storage of the pictorial representation is most complex. There are many variables to consider, beyond the normal variances in our personal ways of drawing things. We must also consider such items as font, line thickness, color, size, etc. Storing chemical diagrams at greatest complexity requires holding:

1. One or more compounds per file.

2. 2-D and 3-D visualization.
3. Other pictorial information (style, color, line thickness, etc.).

One solution to the problem is to do what many have done up to now: ignore the pictorial and concentrate on the chemical. Unfortunately, the pictorial is extremely important to ourselves, and impacts more work lives (secretaries, editors, salespersons) than the other categories. If in fact we decide that our primary end-user is at the institutional, rather than individual, level, then we must allow for the legions of *non-chemists* employed by those firms who rely upon consistent pictures because they do not understand the underlying structure.

We cannot dismiss the importance that we place upon the appearance of a structure, even if we are able to decipher different representations of the same structure. How many of you have ever rejected a slide, not because the data on it was wrong, but because it did not look right? How many of you have ever missed a literature reference when scanning a journal because a structure was drawn in a nonrecognizable form?

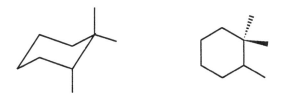

Figure 1. Pictorial representations

Even the coordinate systems which we employ are vulnerable to intellectual prejudice. It may be clear that we need to make provision for storing atomic coordinates in three dimensions. Yet the "standard" esthetically pleasing diagrams which we see in journals often have (X,Y) coordinates on the printed page which cannot be related to their (X,Y,Z) coordinates in "real" space. This is very true of stereochemical representations, where the so-called "real" shape of the molecule is translated into an unreal flat appearance.

Can we say which picture in Figure 1 is correct? Is the correct picture one which more closely approximates the actual shape of the molecule, or the one which allows the reader to more easily discern the molecule being described?

In developing a standard, we will have to decide how to deal with allowing the two representations to coexist, and furthermore, to allow 2-D and 3-D applications to exchange data without destroying the work done by the other type of application. In other words, I should be able to draw a compound for perhaps a grant proposal, then model it, and still have access to either the 2-D or 3-D visualization on demand. This accessibility should be regardless of the order in which the applications are used.

I have encountered many chemists who seem to feel that applications exist in a clear hierarchy defined by their perceptions of which applications are more "important" than others. As an example, in the area of exchanging data between data bases and publications, I have occasionally been asked if a diagram of a compound in some data base can be added to a report. Rarely am I asked if a diagram from a report can be added to a data base. Yet the question presumes that the compound was drawn for the data base before it was needed for a report, although in most organizations, the compound must be drawn for an internal report before it can be entered into a company's data base. Therefore, although data bases have a higher rung on the intellectual ladder, the lowly report writer is the point at which the structure is first drawn. Again we must humble ourselves and remember that the intellectual hierarchy is not necessarily the organizational hierarchy.

So we are left with the question of deciding how much information should be included in a description of a structure. Do we include the vastly more complex (and difficult to maintain) pictorial representation? Can we afford not to? This is a critical decision affecting the size and complexity of the structure files.

Who will define and maintain the standard?

Once we have decided who will be served (or ignored) by a standard, we must proceed to decide how such a standard will be developed and established. We will also have to be prepared to support extensions to any standard, with such extensions typically being either application-specific or vendor-specific.

When the time comes to sit and discuss a standard, it is clear that the primary players in the effort will be vendors. End-users are almost powerless to impose a standard because they are generally incapable of acting in concert to force the vendors to comply.

Because the vendors already exist in the market, they bring with them to the discussions:

1. A bias toward their own products.
2. A "can't do" attitude.
3. A desire to protect proprietary information.

The vendors are motivated to confine the standard to those items which they already support. This is not simply a selfish approach; rather, each individual vendor's constituency is basically satisfied by that vendor's product and may not have a mandate to go beyond it.

Vendors are, of course, experienced in the field, and they bring the important practical knowledge of the current technical limitations. However, experience also tends to be cautious, especially when estimating the difficulty of implementing change. Many managers have known that when sheer speed is required, new employees are best, because they do not yet know that they are being asked to do the impossible. Vendors will be inclined to design within practical, rather than theoretical constraints.

Why would a vendor want to support an emerging standard? Standards have two effects which really bother vendors. First, a standard obviously paves the way for users to switch vendors easily. Second, a standard can constrain applications software. Have you ever wondered why vendors are often reluctant to divulge their file formats? The stock answer is that they do not want to commit to supporting a file format that may change in the future.

However, another powerful reason exists: formats for the storage of information reflect the theoretical capabilities of the software. If I know what data a program saves and how it saves it, I can make reasonable predictions of how the software behaves internally, and then I can predict what types of enhancements will be easy or difficult for that vendor to implement.

End-users are not without their own biases. It is difficult to get adequate representation of the entire end-user community. Since any end-user contribution comes either through contacts with vendors or through voluntary service to professional organizations, it is difficult for people who are not institutionally affiliated to provide input. Therefore, since mankind and individuals (acting solely on their own behalf) are both difficult to represent, the end-user point of view tends to come only from an institutional perspective.

Even the institutional end-user community brings biases because the representatives are current users, and those potential users who are not yet involved are ignored. This is a common fault of product marketers. While there is the saying that a company should always listen to its customers, some companies forget that they also need to listen to those who are *not* their customers.

Therefore, when judging the results of any collaboration, you must critically examine the source. Even if no bias is apparent, can you be sure that the representation was not skewed? As you read my remarks, given in the context of professional discourse, you still must ask yourselves questions such as: Who is this man? What axes does he have to grind? Who is not represented? Why do they not have a forum for speaking?

To this point, I must state that I have had nothing but cordial relations with those working toward a standard. Nonetheless, true representation occurs not just by allowing anyone with an interest to participate, but by proactively seeking out representatives of all affected parties. We must educate those who are not interested, but would be affected, in order that they will become interested once they understand what is at stake.

Lessons from other standards

We are not the first people to try to develop standards. For as long as we have been civilized, our social needs have forced us to develop standards in order to communicate among ourselves. We can learn from the mistakes of others, even those in fields far removed from chemistry.

From personal observation, there are several consistent patterns which emerge whenever anyone in any industry attempts to establish a standard. First, it seems that people want to have a choice. Rather than having a referendum on a single proposed standard, people seem to prefer choosing between two alternatives. It is rare that any standard evolves without competition, especially during developmental stages. A winner sometimes emerges, but that winner is decided in the marketplace, not by committee. Some choices which we have seen in our everyday lives are: Coke vs. Pepsi; Republican vs. Democrat; VHS vs. Beta; Cassette vs. 8-Track; IBM PC vs. Macintosh.

From these examples, and others which are going on today (Compaq Computer is forming a consortium to develop an alternative to the PS/2 bus; several UNIX software vendors have formed the Open Software Foundation (OSF) to combat the AT&T/Sun Unix alliance), there is a lesson to be learned.

Lesson 1. After someone, especially a dominant (or potentially dominant) vendor announces a standard, some "upstarts", perhaps acting in concert, may develop a competing standard.

What this means in practical terms is that the first announced standard may not resolve the conflicts which exist today; it may turn out instead to be viewed as a declaration of war.

Another area where a standard has been established is related to a related topic: the standardized interchange of computer graphics. One attempt at standardization in this area produced the Computer Graphics Metafile (CGM) format. This standard is the work of an ANSI committee. The draft is over 250 pages long, and it allows for an incredible number of ways of describing a computer image. Naturally, despite the committee's desire to be comprehensive, some very basic drawing elements were left out.

The CGM standard is gaining acceptance, but the complexity of the specification has led to a problem of incomplete implementation. There is a real tendency for vendors to ignore the implementation of those elements which are deemed to be so complex that no-one would be using them. Today, among the dozens of software packages which purport to support metafiles, many are incapable of exchanging files because their implementations differ.

Lesson 2. Standards which are so comprehensive that they outstrip current technology will not be implemented consistently across vendors and applications.

It is easy to design a specification that is theoretically comprehensive, but to do so violates two of our original guidlines; namely that vendors must be both able and willing to implement it. This leads to a corollary: Despite a written specification, a substandard often emerges after several rounds of vendors and users have been forced to digest the incompatibilities. The substandard becomes the *de facto* standard.

Many, if not most, of you will own an IBM PC clone. Today almost every clone is billed as being 100% IBM compatible. Yet no clone is actually 100% compatible, because if it is, it will violate IBM's copyrights and patents. For example, IBM's original BASICA program will not run on anything other than a true blue IBM PC. In the early days of PCs, software vendors took advantage of every hidden nook and cranny of the IBM PC, and clonemakers were unable to fully replicate everything. The early clones found themselves unable to run many different programs. The problem has disappeared today, but only because both software vendors and PC clonemakers had an incentive not to push adherence to the "standard" of the IBM PC beyond a reasonable point.

Therefore, it is reasonable to assume that any developing standard for chemical structures will eventually develop vestigial parameters which are never used, and certain extensions to the standard will become integral to the standard.

Another area we can learn from is the exchange of text documents. How many of you have ever tried to exchange a word processing file from one program to another? How many succeeded, while still maintaining intricate formatting details? What happened if you tried to get it back from the other program?

In text processing there is an IBM protocol known as Document Content Architecture, or DCA. Virtually every major word processor, including Word Perfect, Microsoft Word, and Displaywrite 4 claim to export and import documents through DCA. If you have any two of these programs, beware of trying them on anything more complex than a one-paragraph memo, and even that may not emerge unscathed.

Lesson 3. The one-way transmission of data is far simpler than a smooth two-way exchange. Despite the existence of a "standard", it is rare that data can be exchanged by diverse applications without the loss of at least some auxiliary data.

As discussed earlier, in the case of chemical structure data, if you pass a molecule from a 3-D package to a 2-D package, even if you perform no manipulations in 2-D mode, the mere act of converting the data can result in a loss of the Z-coordinate data. If the eventual standard includes pictorial data, the danger of data loss increases dramatically.

Finally, we must realize that we cannot predict the future, and that new developments may well render current efforts obsolete. As time passes, a standard which was once liberating becomes constraining; the symbol of progress and cooperation becomes an ankle weight dragging down those who slavishly adhere to it. Since standards reflect the applications, they eventually constrain applications which adhere to the standard.

Will a standard solve our problems? Of course not. As applications progress, some intrepid vendor will extend the standard to accommodate a new advance. The end-user will then have to evaluate whether this breach of exchangeability is worth the additional benefits of the new software. Progress often requires the violation of standards.

Standards happen last, not first. To wait for a standard condemns a user to being the last to convert to new technologies. No matter what happens in the arena of standardization, you will not be relieved of your obligation to THINK!

RECEIVED May 2, 1989

Chapter 3

Information Integration in an Incompatible World

Dennis H. Smith
Molecular Design Ltd., 2132 Farallon Drive, San Leandro, CA 94577

Computer-based information on chemical structures encompasses several levels of descriptions of the structures themselves (internal data structures, external files, graphical forms), together with a wide range of associated numerical and textual data. This collected body of information must be available throughout an organization. Special requirements, for example research, or regulatory affairs, will dictate which portions of the information are necessary for a particular end use. Sharing of this information requires systems that are compatible at some well defined level of information exchange. Although all agree that such compatibility is essential, we must understand the powerful, interrelated forces at play that restrict compatibility, including human and dollar costs of retraining, absence of system hardware and software standards, a highly competitive marketplace, and rapidly changing technology. These forces must be understood in order to plan for the chemical information systems of the future.

Information has been characterized in the popular press as a *weapon* which can be used to gain a strategic, competitive advantage. This is certainly true in the chemical and pharmaceutical industry and is, in my opinion, true also for chemical research taking place in the academic and not-for-profit sectors. I define chemical information in a very broad sense, including structural information on chemical substances, together with associated numerical, textual and graphical information. The discovery and rapid delivery of important chemical information from computations, or in notebooks, company reports or the open literature can dramatically accelerate the progress of research and development on new chemical or biological methods and products.

Virtually all organizations have turned to computer systems to manage their information. Investments in hardware, operating systems, and application software have been extremely high, as organizations seek ways to exercise the information weapon. During this period of focus on computers and information, the world of computing has been changing dramatically. Several changes are especially notable. Firstly, the price to performance ratio of computer hardware has declined radically, and we can all anticipate a very powerful computer on our desks if we do not already possess one. Secondly, the ratio of software to hardware costs is increasing, leading consumers to ask harder questions about software integration and quality.

0097–6156/89/0400–0018$06.75/0

Thirdly, software advances are lagging advances in hardware substantially, making it more difficult to take quick advantage of new generations of hardware. Fourthly, application software has previously emphasized functionality over integration. Fifthly, application software systems will be required to integrate or share information with other systems in the future. How this is achieved is not important to the consumer, it simply must be done in order that they can take advantage of available information while trying to manage rapid change.

Achieving true information integration while the computer industry is undergoing rapid change is a major challenge. To explore the limitations and the possibilities for the future, I have divided this paper into three parts. Firstly, I discuss some of the general trends in the computer industry. Secondly, I discuss how these trends affect creation and delivery of chemical information, with a focus on chemical structures. Thirdly, I discuss bad news and good news about the future.

General Trends in the Computer Industry

A computerized information system designed to promote integration of information across an organization (the term enterprise-wide computer systems has been used by many authors) will be based on hardware and software systems, which I divide into three elements: (1) hardware; (2) operating systems and environments; and (3) application software. All three elements must work smoothly together in order to have a useful system.

Although I am anticipating some of my conclusions, the discussion below reveals three important facts. Firstly, each element of a computer system is provided by a few to a few hundred companies. This produces an obvious barrier to integration, since no one company provides all elements of an information system in an acceptable form. Secondly, these elements of computer systems are maturing at vastly different rates. For example, hardware developments dramatically outpace application software developments. At the very least, this slows a consumer's ability to take advantage of the latest advances in hardware. More often, incompatibilities are introduced that prevent use of new hardware. Thirdly, although many advantages have accrued to consumers of computer systems, many of the advantages carry with them the defects of their virtues. In other words, progress results in simultaneous creation of both advantages and disadvantages for any given advance.

Hardware: the Revolution. Everyone agrees that a revolution is taking place in computer hardware. It is clear that we can, if we choose, have workstations on our desks in the 5-20 MIP range, at today's personal computer prices, in the next few years. Some moderately priced machines achieve the lower end of that scale today. But this revolution is not limited to raw cpu power. It extends to available memory, computer networking, hard disks, specialized processors, etc.

Advantages. This change in hardware is changing the way we think about computing. Machine cycles will no longer be an issue; this will free creative people to worry about other aspects of information integration. Computers will become commodities which are purchased incrementally from a variety of vendors to meet increased needs for access to machines. These advances will be driven by the coming generation of distributed computing environments, employing a mix of workstations and shared file and compute servers.

The ready availability of powerful, inexpensive machines has dramatically broadened the profession of programming and the pool of talented programmers. The explosion in the number of software companies, some of them very successful businesses, is evidence of this. Many useful software tools have been developed, driven by cheap hardware for both the software developer and the consumer.

Engineers of hardware systems, taking advantage of dramatic improvements in systems for computer-aided design, can rapidly develop special hardware for special application. There are many companies providing custom chips, and many companies working on a new generation of minisupercomputers.

Disadvantages. The rate of change is far greater than our ability to assimilate the machines that result. Although some machines, especially very high speed processors, are developed to solve special problems, for the most part we have not clearly defined the problems that require such speeds. However, the increased speed of standard workstations is partially consumed by the new generation of operating environments (see next section). Many organizations are still trying to find useful applications for personal computers beyond use as (high priced) terminals that use software for terminal emulation. Other companies are still working with terminals connected to mainframes. The former group may adapt readily to distributed systems, but the latter group faces a large capital investment.

The consumer of machines is confused by the plethora of choices of basic machines of several flavors and configurations, linked together with a variety of networks, and using a variety of peripherals from a thousand different vendors. Confusion increases every day, with every new announcement.

The lack of clear solutions to clear problems, and the confusion of choices leaves most consumers in a quandary. Even if new hardware can be justified, which should I buy? Which of today's new hardware announcements will allow me to integrate my existing, heterogeneous computing environment?

Finally, computers today are *not* commodities. Hardware incompatibilities abound, all the way from cables and mice, through cpus and expanded or extended memory, to networks. Only a few of the incompatibilities can be masked completely by software.

Operating Environments (Systems): the Reform. Although the definition and function of operating systems is clear to many people, I extend this definition to the new generation of operating environments (OEs) that will be supported within distributed computing environments. The operating environment is the generic interface to the system as supplied by the vendor of the operating system. The OE includes the operating system and the manner in which an interface to the operating system is presented to the end-user. For example, in the DOS world, the OE is an ASCII terminal with a DOS prompt, unless someone else has provided additional functionality. In the Macintosh world, the mouse-driven, windowed operating environment shields the user from the operating system itself.

Operating systems and environments are also in a rapid state of change, although they are maturing far more slowly than the hardware platforms on which they run.

At the same time, they are advancing far more rapidly than application software, so I place them in the intermediate category of "reform."

Powerful operating environments are already available. Most are in the general category of "WIMPs", Windowed, Icon-based, Mouse-driven, Pointing systems. Developers of application software are moving their applications into these environments rapidly, to take advantage of the benefits that result.

Advantages. New developments in environments are driven by a recognition that information must be shared among people, many of whom have little knowledge of, or experience with, computers. A well-designed operating environment can reduce training costs and increase productivity. A well-designed interface that makes effective use of multiple windows, graphics, and multitasking, if available, is simply much easier and more fun to use.

The world is slowly standardizing on a limited number of operating systems. Rather than introducing new operating systems, vendors of new hardware are attempting to offer an existing, popular operating system. This is a tremendous benefit for both the software developer and the consumer.

The same operating systems and environments, for example UNIX and X-Windows are available for a variety of different hardware platforms. Some vendors, for example, Microsoft Corporation provide very similar environments (Microsoft Windows series) on different platforms. This frees the consumer somewhat in his or her choice of hardware. The end-user will see a similar environment on a number of different hardware platforms.

Disadvantages. The first disadvantage in operating systems and environments is that there are *still* too many of them. Most hardware will support, or is offered with, only a restricted number of operating systems and environments, often one. Thus, the consumer is not yet free to choose among hardware in order to obtain a given environment. Of course, the operating systems and environments are not compatible with one another, nor do they offer the same look and feel to the end-user. Integration of information across heterogeneous machines and operating systems is currently impeded or impossible.

The second disadvantage is that the functionality of good operating environments does not come for free. Good OEs all require a substantial machine to run them effectively (efficiently). On machines on which they run well, they consume a significant fraction of the increased cpu power being generated by the improvements in hardware. In some applications, the net speed improvement for the end-user is very small, even though the user has a more powerful machine.

Application Software: the Evolution

As most people know, the rate of development of new application software is improving, but is still dramatically slower than the rate of development of new hardware. The advent of structured programming, better software engineering practices, and some recently released tools for computer-aided software engineering, are all helping. The relatively small number of popular programming languages and operating systems makes it easier (but not easy) to port applications to different machines. The process of producing well-integrated application software is an evolutionary one, not a revolutionary one.

The chemistry community requires software systems that integrate information from several different sources, for example, structural information with numerical data with textual data. Systems that manage or analyze those data are provided by many different vendors, and often run on different hardware platforms in different operating systems and environments. Information integration under these circumstances is a real challenge.

Yet many efforts are underway, through formation of joint development groups, sharing of file formats, and creation of customized and callable programs. Virtually all of this work is at the interface level of information exchange. Few software firms actually share code.

Advantages. Any improvements to the process of software development, or the ability of different third-party providers to exchange information will obviously benefit the consumer. There will be more examples of better integrated software, however, as developers learn how to work together to provide complementary solutions.

The availability of low cost, high performance hardware has made possible the development of many thousands of software packages across a broad spectrum of applications. Even though the productivity of each individual software developer has not increased substantially, the fact that there are now so many developers has given us many, diverse products at relatively low cost.

Disadvantages. Given that the software industry will remain highly fragmented, most information integration will have to be derived from two or more companies working together. The problem is that most software companies have many more customers for their stand-alone products than customers that require the smooth integration of two or more products from different vendors. Managing compatibility given asynchronous development and release schedules among vendors is very difficult.

The quality of application software will receive increasing attention from consumers in the future. A significant amount of software being produced today is being produced by individuals who are not sufficiently trained in, or motivated by, requirements for quality. Quality is a critical issue for the consumer, whose business may depend on the results produced by an information system. Several factors make obtaining high quality software a difficult task for the consumer: (1) The fragmentation of the software industry leads to many, slightly different products to perform the same function. A consumer seldom has the time for the careful analysis required to determine what software package produces the most accurate results. (2) Software quality assurance is a fledgling discipline. There are relatively few trained people, and software testing remains an inexact science. (3) The highly competitive nature of the marketplace and the bottlenecks introduced by thorough testing often force smaller companies to deemphasize quality in order to get products shipped.

Standards: the Holy Grail. "My standards are better than your standards." (Anon.) Consumers are becoming increasingly aware of the incompatibilities introduced by a fragmented computer industry. Even with this awareness, they are often forced to choose multiple vendors in order to get solutions to their particular problems. In this environment, computer and application software vendors have begun adopting standards as a way of reducing incompatibilities. Results to date are mixed. Standards have emerged, and will continue to emerge. They will certainly help resolve incompatibilities. But they will not solve all the problems quickly.

Hardware Standards. Although (1) hardware will become more of a commodity, (2) some hardware standards already exist, and (3) other hardware standards are rapidly being adopted, each hardware manufacturer is under powerful pressures to maintain product lines that are strongly differentiated from those of its competitors. The growth and profitability of the major computer manufacturers is still driven by selling hardware. These manufacturers are struggling to determine how to remain in business in the future as profit margins on hardware continue to drop and as pressure builds for more compatible, and thereby less differentiated machines.

Different vendors are taking different approaches. For example, Digital Equipment Corporation (DEC) and International Business Machines Corporation (IBM) are continuing to produce proprietary, and incompatible, processors and networks, although third-party manufacturers of peripherals do offer compatible equipment. Apple Computer is jealously guarding its proprietary rights to the Macintosh architecture, although the recent announcement of a relationship with DEC will improve network connectivity between their respective machines. On the other hand, Sun Microsystems (SUN) has clearly made the decision that it is in their best interest to license aggressively its SPARC architecture, which will enable other manufacturers to build compatible machines.

Several chip manufacturers, for example Intel and Motorola and many other, small producers of advanced cpus, make their own processor chip sets and sell them to computer manufacturers. This approach has led to the construction of a large number of compatible machines (clones) by different manufacturers. This has promoted information exchange dramatically, but done little for true information integration across heterogeneous systems. In addition, the chip manufacturers are themselves in fierce competition producing proprietary and incompatible chip sets.

Operating System and Environment Standards. Probably the most dramatic examples of the wars over standards and compatibility are found in the area of operating systems and environments. All computer manufacturers provide as their primary offerings, proprietary, and incompatible, operating systems. For example, IBM offers several operating systems across its line of computers and is expending substantial resources in its Systems Applications Architecture project to ensure compatibility at some level among its own offerings. DEC offers VMS which is compatible across its VAX line of computers. Apple has made its position very clear. Apple regards its operating environment for the Macintosh as proprietary and of critical strategic importance to the future success of Apple. None of these positions promotes information integration.

The UNIX operating system in several variants is now offered by virtually every computer manufacturer, including all those mentioned in the previous paragraph. Although the operating system is regarded by many as arcane for the end-user, and the availability of application software in the area of chemical information is much more limited than for other operating systems, UNIX at least offers some hope for more portable software. Its availability makes it much easier for a consumer with a UNIX-based application to choose the best hardware platform to support the application.

Standards are emerging for operating environments. For example, Microsoft Windows and its close relatives, including Windows/286, Windows/386 and the Presentation Manager, provide an operating environment that is similar across many different machine in the MS-DOS and OS/2 worlds. X-Windows is emerging as a standard in the UNIX and VMS worlds. Hewlett- Packard (HP) has announced New Wave, an operating environment based on the Microsoft Windows architecture. These environments are incompatible, but they at least offer some similar functionality and appearance to the end-user.

A closer examination of the efforts toward standards reveals, however, that competitive forces similar to those noted above for hardware are found in operating environments as well. This is not surprising, given that the OE is usually supplied by the computer manufacturer. Thus, the OE becomes a key element of protection of a proprietary system and in differentiating one system from another. None of this promotes information integration across heterogeneous systems.

The vendors have carried the competition one step further. Recently, in an effort to protect their positions, several companies have resorted to the classic defenses of litigation and obfuscation:

Litigation. For example, Apple, in attempting to protect its investment in the Macintosh and its OE, has sued both Hewlett-Packard and Microsoft. The former is charged with copying the look and feel of the Macintosh OE in its New Wave architecture. The latter is accused of violating an earlier agreement between the companies on sharing OE technology. Recent court decisions have lent some credence to the argument that "look and feel" can be copyrighted, so these suits must be taken seriously. A software company writing applications for either the HP or Microsoft OE is obviously given some cause for concern. This situation does not promote standards of compatibility or consistent OEs.

Obfuscation. For example, consider the UNIX wars. UNIX is an "open" standard (see below), and for years several different versions have been available. Vendors have been under pressure to support a single version. American Telephone and Telegraph (AT&T) and SUN recently announced their intentions to build and support a single version of UNIX. With motives that have been questioned by some observers, many other computer vendors have responded by joining together as the Open Software Foundation (OSF) to produce their own "standard" version of UNIX. This is, of course, nonsense, and destroys the concept of standards. A software vendor would have to build for, and test extensively in, both OEs in order to deliver a quality product.

Additionally, we should not be surprised to learn that some of the standards being promulgated by the vendors in an attempt to sell systems are not in fact standards at all. Many are in the form of "extendable" or "open" standards, both of which are oxymorons. Classic examples are X-Windows and UNIX:

X-Windows. X-Windows is a misnomer. It is actually a network protocol and "knows" very little about windows. X-Windows is under development by Project Athena at MIT, and has been made available in a series of releases to members of the industrial consortium connected with the Project, and to anyone else desiring a copy. The problem is that everyone gets source code, and of course the temptation to modify and enhance it cannot be overcome.

The fact that the windowing functions provided by the base level implementation of X-Windows are exceedingly primitive means that it is up to "toolkits" and application software actually to produce a functional operating environment. Until very recently, it was the responsibility of each computer vendor to supply its own toolkit, and as expected the toolkits have been incompatible. While this paper was under review, the OSF has established standards for the operating environment on machines produced by its members. This standard adopts: (1) the "look and feel" of the Hewlett- Packard/Microsoft Windows/Presentation Manager interface; and (2) the toolkit provided by Digital Equipment Corporation. This decision is a significant step toward standardization although it leaves Sun Microsystems and AT&T still with a different operating environment. This diminishes the principal concept of a standard, that it actually be standard in all its manifestations. The burden on companies that try to produce the same "look and feel" of an interface under different versions of X-Windows will be considerable.

UNIX. The difficulties with the UNIX standard were introduced in the previous section. The fact that two major (Berkeley and AT&T) versions, plus several derivatives, have been available has seriously impeded the porting of major software systems to hardware platforms supporting UNIX. This is especially true for applications requiring a highly interactive, graphical interface. For several reasons, especially performance, OE's that support high quality windowing systems have been forced to modify UNIX. The current wars over the UNIX "standard" create enormous confusion among developers and consumers alike, and seriously diminish the effectiveness of a significant advance in OEs that could promote information integration.

Application Software Standards. There are several standards in place or evolving for application software in the area of managing information. These standards are being driven largely by consumers who require the integration of information from several different software systems, each produced by a different vendor for a specific purpose. Vendors are now responding to these requirements and are beginning to provide more software that adheres to standards. Many standards are at the level of information exchange between or among programs, using compatible files and/or utility programs, for example, the Digital Document Interchange Format (DDIF) for the ODA standard (below). Other standards, for example SQL (below), are at the level of description of command syntax that allows access to data stored in assumed ways. Appropriately written application software can provide, using these standards, direct access to the data which may have been created in, and stored by, other programs.

Some standards have the backing of an appropriately constituted standards committee, such as the American National Standards Institute (ANSI). Other standards are in the category of "open." It is instructive to contrast the two different categories for their effectiveness. Here are some illustrative examples:

SQL. SQL is the Structured Query Language standard promulgated by IBM. It is the *de facto* standard for relational data base management systems (RDBMSs). The standard defines the syntax of commands that can be used to access and retrieve information from data stored in one or more "flat" two-dimensional tables. Most

135,057

vendors of DBMSs provide a "host language" or "application program" interface to their systems, allowing other software to "call" their access and retrieval code directly. The syntax of the queries is controlled by the SQL standard. SQL has proven to be an extremely effective standard in the DBMS world and is provided by virtually all vendors of DBMSs. One drawback is that the standard itself is extremely limited, and most vendors have added their own extensions. This inhibits compatibility, and is an irritant to the end-user, because third-party application software which adheres strictly to the standard may have diminished functionality compared to the vendor-supplied access to the database. So far the advantages of the standard have outweighed these incompatibilities.

ODA. The CCITT (Consultative Committee on International Telephony and Telegraphy) X.409 standard defines the Office Document Architecture (ODA). ODA is an emerging standard for interchange of "compound" documents, i.e., documents which contain mixed text, data and graphics. Today, such documents are produced in a large number of word processing and desktop publishing systems, most of which are incompatible. The ODA standard will help this situation considerably, assuming that vendors endorse and support the standard. However, several vendors have already expressed their displeasure at the graphics library supported within the ODA standard, and plan to supply a different library. Only time will tell if the base standard proves successful for promoting document interchange.

PostScript. PostScript is a page description language developed by Adobe Systems Incorporated. Today, most word processors and desktop publishing systems support PostScript as one of their output formats. Most laser printers and many document production systems support PostScript as input. An extension of PostScript, called "encapsulated" PostScript, creates an output format which can be imported into some document production systems as a way of integrating information from different systems. PostScript is not strictly an interchange format, since the imported material in encapsulated PostScript cannot be edited, merely printed. This may change in the future.

PostScript, in combination with laser printers, has really revolutionized document production. It is noteworthy that PostScript is owned and controlled by Adobe Systems and is licensed to other hardware and software companies. It is not "open", it is stable, and it is very effective.

TIFF. The Tagged Image File Format was an early "standard" for the exchange of compound documents. It is an "extendable standard" and the fact that many of its supporters have chosen to do so has made document interchange at best difficult or incomplete, or at worst, effectively impossible. TIFF has not been a success as a standard to the same extent as has PostScript.

Universal standards. This term may also be an oxymoron in today's computer industry, but there have been some successes in defining standards that will enjoy broad acceptance. For example, the definition of the Open Systems Interconnect (OSI) networking standard has promoted standardization and will foster more compatibility among networks. There are a number internationally recognized organizations that are working on standards in many areas. These include the International Standards Organization (ISO), ANSI, and the Institute of Electrical and Electronics Engineers, Inc. (IEEE).

One way to approach standardization for information exchange is to define an intermediate, common data structure or file format to which a variety of incompatible formats can be converted, usually through a utility. A companion utility is used to convert from the common format to a format compatible with another system. A good example of this approach is the utility or interchange format offered by Keyword Office Technologies, Ltd., to allow interchange of text from various (incompatible) word processors (WPs). This approach reduces an incompatibility problem, given n incompatible word processors, of order n^2 to a much smaller function of n, as illustrated in Figure 1.

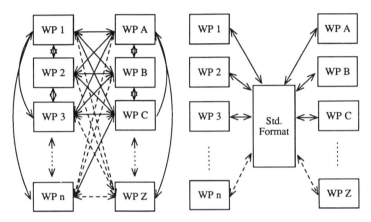

Figure 1. (left) The order n^2 problem if every word processor (WP) format must be compatible with every other WP. (right) The order n problem if one must interchange among n WPs.

As illustrated in Figure 1, left, one can direct exchange formats between two WPs. If this is done carefully, one can generally ensure compatibility. However, bidirectional, direct conversion implies $n(n-1)$ different conversion utilities. An intermediate, standard format (Figure 1, right) simplifies the problem, in that one requires only one (bidirectional) utility per pair of WPs.

A significant problem, however, is introduced with the concept of the universal standard interchange format (for example, the "Std. Format" of Figure 1). The interchange format must be a superset of *all* the formats that it interchanges. In other words, the utility program that performs the interchange must understand the respective features, or functionality, of its source and destination formats. For WPs, the Keyword utility attempts to convert 100% of the text, emphasis, headers, footers, sub and superscripting, etc., of one WP, through its interchange format to the corresponding capabilities of the target WP. Obviously, it is very difficult to maintain 100% compatibility, especially with the rate of change of, and enormous differences among, WPs in today's market. The problem of compatibility may actually end up being worse for any given pair of formats, if incompatibilities are introduced going both to and from the interchange format. There is a direct analogy with the interchange of information on chemical structures, as described below.

Implications for Chemical Information

Turning now to the narrower area of use of computers in chemical information, how do recent advances in hardware, operating environments and software, together with standards, affect the discipline of chemistry? To the extent we rely upon the computer industry to supply products, the answer is that almost all the advantages and disadvantages discussed above accrue to chemistry as well.

Hardware. There is one common thread to use of computers in chemistry: virtually everyone is using them. Beyond that, it is clear that computer hardware has been purchased to solve problems specific to individuals, or a laboratory, or a department or division. Issues of compatibility, integration, networking, etc., were often ignored, and still may be ignored if a particular platform is best suited to solve a particular problem. Most organizations have an enormous capital investment in machines that often cannot be connected, or can be coupled only loosely for file transfer (information exchange).

Although most organizations in the chemical industry have IBM or DEC mainframes, our information for the industrial sector is that access to these machines is still largely through terminals. As workstations (and I consider high end Macintosh and PC AT class machines to be workstations) are becoming more widely available, we note different buying patterns around the world. If there is any generalization, it is that no standardization on hardware will occur.

People will have hardware from DEC, IBM, Apple, HP, SUN, etc., etc. This diversity of hardware, much of it incompatible at all levels of information exchange, let alone integration, creates chaos and confusion in our community as well, but that is today's reality.

Operating Systems and Environments. Most people would like a single, consistent interface to all their chemical information, including structures, data, images and text. The first barrier to providing consistency is the variety of operating systems in use in the community. Some application software (see below) has been written to shield the end-user from the operating system, and to be portable across machines and operating systems from different vendors. Other software remains locked to a particular piece of hardware and accompanying operating system.

The advent of operating environments that are fewer in number than the number of operating systems will improve the situation. I expect that the chemical community will follow the computer industry and "standardize" on a small number of OEs. I expect that the following OEs will find favor in the community, and be chosen by developers as the platforms on which to build the next generation of software for chemical applications: (1) Microsoft Windows for high speed 80286 and 80386 DOS machines; (2) Windows' cousin, the Presentation Manager, for 80286 and 80386 OS/2 machines; (3) Finder and Multi-Finder for the Macintosh; (4) X-Windows and toolkits for UNIX machines; and (5) DECWindows for DEC/VAX/VMS.

For the software developer, the fact that all environments have related functionality is a step forward. For the end-user, it is arguable whether an advantage will be achieved. Users want both (1) a consistent interface across platforms, and (2) preservation of platform-specific idiosyncracies so that different applications on the same platform provide similar look and feel. Each OE imposes a particular style of interaction with the machine that cannot, and probably should not, be shielded from the end-user by an application program. It remains a formidable challenge to support two or more OEs from a single application program in a way that is sensible and acceptable to the end-user. However, we think that such interfaces will be available in the near future as software vendors confront the reality of a highly heterogeneous workstation environment.

Application Software. Although there is a wide variety of software available to the chemical community, each package has been built to solve a particular set of problems, and until recently, information integration has been difficult. Customer requirements have led some companies to work together to provide new capabilities for information integration or exchange. For example, at MDL we have formed several strategic relationships with other suppliers of hardware, OEs and application software to the community, including, for example, IBM, Hewlett-Packard, DEC, Interleaf, Inc., and Oracle Corporation. In this way we can provide more integrated chemical information systems, systems that take advantage of the strengths of the complementary products of the various vendors. Other companies in the industry are adopting similar approaches.

These are steps in the right direction for consumers, but much more remains to be done. Each software provider has its own interface to data, and integration of two or more programs inevitably leads to inconsistencies in interfaces, or a different mix of functionalities, either of which makes it difficult for the end-user. Expanding the scope of strategic relationships will provide better integrated solutions. However, there are many gaps in information integration that remain to be solved, due to several factors:

Businesses, to be successful, must make and sell a useful product in the midst of intense competition. In a vertical, and limited, market such as the chemical community, similar products must be differentiated in order to attract customers. This fact spawns a large number of small vendors providing similar products, for example, in the area of PC-based molecular modeling software. The products are usually incompatible and have different user interfaces.

Strategic relationships are most simple to form between vendors who have complementary products. It is unreasonable to expect cooperation between two vendors who are in competition with one another. Although such cooperation may benefit the consumer, it is usually poor business practice, and may jeopardize the existence of one or both of the vendors. This fact contributes to perpetuating incompatibilities once they arise. A good example of this factor is the number of different user interfaces to chemical data and structures. Consumers would like one interface, but no one vendor provides access to all information. Each vendor has invested a large amount of money in developing its interfaces, and its customers have invested a large amount of money in training their respective end-users to use the interfaces.

There are very poor links between online scientific and business information on the one hand, and research laboratory, or industrial proprietary chemical information on the other. Integration of this complementary information is difficult.

Different methods for managing structures and data in different software systems prevent information integration. These differences go far beyond the user interface, and can lead to fundamental incompatibilities which are difficult if not impossible to overcome. The best examples of this are differences in representing chemical structures, discussed in more detail below.

Some consumers now have requirements for integration of documents, text, and graphics, or images with the rest of their chemical information. The community is still struggling with individual standards for chemical structures, for data, and for documents. Standards for integration are not even being considered. Unless we are careful, this will lead to another round of software systems to solve today's problems, and another round of incompatible systems in the future.

Standards – Representation of Chemical Structures as an Illustrative Example. All of the issues about hardware, OEs, software, and standards discussed previously in general terms pertain to the chemical community as well. Rather than trying to discuss the issues in the context of specific examples drawn from chemistry, I will pick a single example and discuss it in detail to illustrate some of the challenges we face.

Nothing is more fundamental to the theory and practice of chemistry than the chemical structure. Virtually all chemical information systems use the chemical structure as the common, or "linking" data type. Yet one of the principal barriers to information integration is the fact that different systems use different methods to represent chemical structures. These methods are generally, on the surface, compatible. But incompatibilities exist at deeper levels, hidden from the end-user. These incompatibilities make two important uses of chemical structure information *impossible* to achieve with 100% precision and accuracy:

Information exchange. It is not possible to exchange structural information between or among systems with both 100% retention of information content and 0% errors. One contributing factor is that not all systems store the same chemical information, making complete and accurate exchange impossible. For example, some systems represent stereochemistry, others do not.

But errors can also result in exchange of information between systems whose chemical representations seem on the surface to be quite similar. These errors result from the indeterminacy of translating structural types that are difficult to represent (for example, organometallics, tautomers) precisely in the computer.

A 99% success rate sounds great, until you have 10^6 structures to convert. A 1% failure rate is 10,000 structures!

Structure and substructure search among systems. It is not possible to formulate queries in one system that can be guaranteed to be answered 100% accurately and precisely in another system, for exactly the same reasons it is difficult to convert structures from one system to another. Simple queries may be answered correctly. Complex queries may not be. Answer sets derived from two or more different query systems posed against the same database may each be correct from the standpoint of each query, but be different sets.

How can this be so? For a large number of common, garden variety, classical organic structures, there are few if any problems. Unfortunately, even this class of chemical structures has its idiosyncrasies. When we consider: (1) the ways in which structures and queries are processed and (2) the enormous variety of chemical substances under investigation in the chemical community, the problems, and the answers to the question of how this can be so become much clearer.

Levels of Representation in the Computer. There are many different steps in the storage, retrieval and display of chemical structures. Incompatibilities among systems can arise at any step. These incompatibilities run the gamut from physical data formats through chemical perception of computer representations of structures. Consider the various levels of representation summarized below and illustrated in Figure 2.

Mass storage. Information on disk or tape consists of bits and bytes organized into blocks (and tracks and sectors). Physical data formats on mass storage media

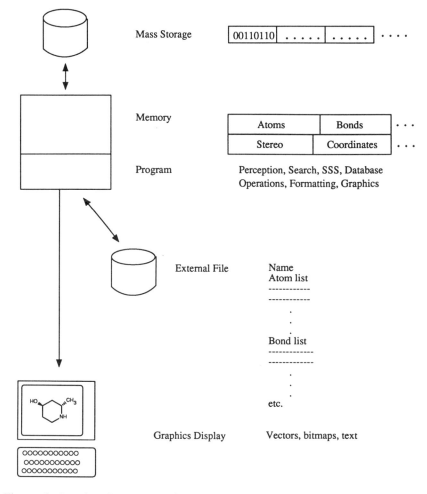

Figure 2. Levels of representation of chemical structure information in the computer.

differ from machine to machine in very fundamental ways, including different word, byte, and bit orders. Chemistry is only implicit in these data, and no system beyond that which wrote the data can read the files and retain the chemical significance without a detailed specification of the format.

Memory. Moving data into and out of memory from disk is generally a function provided by an operating system. A chemical structure is either created or saved by this operation, into or from an internal form in memory that can be interpreted, or "perceived" by a program. The process of moving data in and out of memory from local disks does not itself create incompatibilities. If, however, data are moved between heterogeneous machines *via* tape or network, then incompatibilities can easily arise. For example, the ASCII square brackets "[ ]", and other ASCII characters, may not be supported in EBCDIC. If this is not taken into account, information can be modified or lost in data exchange among heterogeneous machines.

Perception and manipulation. Internal data structures in memory are perceived, as chemical entities, and manipulated by a program. Knowledge of, and rules about, chemistry are intertwined in this process. If different systems choose to represent chemistry in different ways, then the perception and manipulation routines will be different (see examples below), and results of structure exchange, or search will be different.

External files. The primary methods for exchange of chemical structures among programs are: (1) an external file; or (2) a data structure to which the chemical structure is transformed for transmission *via* computer network. The information exchanged is obviously not a copy of a program's internal data structures. What is not so obvious is that the information in the file is usually an interpretation of the internal data structures, involving some degree of chemical perception; it will be transformed in subtle but important ways. For example, one system will have a formalism for transforming an internal representation of aromatic systems, or tautomeric bonds, into the corresponding lists of atoms and bonds as they are written to the external file. Another system, with a different formalism for treating aromaticity and tautomerism, will interpret the external file differently as it is read in and transformed into its own internal representation. This interpretation may or may not yield the same structure.

Graphical objects. A second result of programs for perception, search, etc. (Figure 2) is a graphical image of the structures. This is also derivative information, and a drawing may or may not reflect subtleties of the representation that produced it. For example, aromatic bonds may be drawn as alternating double and single bonds, with no visual cues as to the differences with normal single and double bonds, or they may be drawn as closed circles within aromatic rings, where there is no concept of a "circle" bond in the internal representation. Obviously, similar drawings from different systems may not reflect dissimilarities of internal representations, and dissimilarities in drawings from different systems may result from very similar internal representations.

In the next two sections, I examine some of the issues in chemical representation, and demonstrate by example problems that inhibit smooth integration of chemical structures among systems.

Integration of Chemical Structural Information. Even for relatively simple structures, there are disagreements among chemists and differences among information systems on some fundamental aspects of representation. Consider some of the problems in chemical representation for "classical" organic chemical structures:

Stereochemistry. Some systems represent stereochemistry explicitly throughout all levels of Figure 2; some represent it only graphically, or only with text descriptors. Some systems do not treat the stereochemistry of double bonds. Few systems treat noncarbon stereochemistry. Some systems allow structure and substructure searching with stereochemistry in query structures, others do not. Different systems that represent stereochemistry have different ways of handling relative and absolute stereochemistry. No systems perceive the stereochemistry implicit in the biphenyl system **1**, although it can be represented graphically.

R_1, R_3, R_4, R_2, **1**

Structure 1

Correct perception of stereochemistry is crucial to the use of computers in understanding chemical and biological processes. Given the current disparities of perception of stereochemistry among various systems, it is hard to understand how true integration of this essential chemical information can be achieved.

Aromaticity. Some systems represent, internally to the computer, aromaticity as a property of bonds, other systems represent it as a property of atoms. The bond property may be associated with skeletal, single bonds, or it may be associated with explicit alternating double and single bonds. These alternative representations may be displayed graphically in a variety of forms, as mentioned above. The definition of aromaticity itself differs from one system to another. Such differences can create substantial problems in converting from one format to another. For example, consider the potential incompatibilities raised by the suite of simple structures **2–4**.

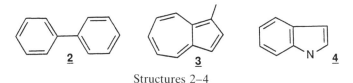

2   **3**   N   **4**

Structures 2–4

Biphenyl, **2**, seems straightforward. There are two six-membered aromatic rings joined by a single bond. What happens to the definition of that bond in converting it to another representation is, however, problematic. A system that perceives the aromatic nature of bonds based on an atom-centered definition of aromaticity may, or may not, perceive the single bond as aromatic during conversion or substructure search. The result will depend on the representation itself and the intelligence of the program that perceives the representation.

The substituted azulene **3** possesses Kekul resonance forms, but would not be perceived as aromatic in a system that restricted aromaticity to six-membered rings. Indole **4** possesses a double bond in the five-membered ring that displays substantial aromatic character in its chemical reactions, yet this ring in indole would not be perceived as aromatic by most systems. As soon as a system introduces fuzzy chemical concepts, such as a definition of aromaticity based on chemical reactivity, into a system based on graph-theoretic concepts, such as perception of Kekul resonance forms, incompatibilities and errors, conversion and searching will result.

Tautomerism. Tautomerism is treated differently from system to system, if it is treated at all. Systems that allow for tautomerism often represent a single tautomeric form, with rule- or table-driven procedures designed to detect tautomeric forms. Again, because there are disagreements on if and how to represent tautomers, exchange of structures between systems is compromised. For example, consider the pairs of structures **5, 6** and **7, 8**, and the reaction from **9** to **11**.

Structures 5–11

The pair of hexenes **5** and **6** does not interconvert under normal conditions of temperature and pressure. However, the pair of hexenones **7** and **8** is tautomeric, and representations and corresponding search systems must take this into account in order to guarantee finding one when querying for the other. Many systems cannot do this because either they do not consider tautomers or they do not allow carbon atoms in the tautomeric system involving labile atoms and bonds.

Finally, the tautomers **9** and **10** constitute a pair of structures where the formalism for representing aromaticity is intertwined with the formalism for representing tautomers. Cytosine, which can exist in two tautomeric forms **9** and **10**, must be in the form **10**, which formally disrupts the aromatic system, to yield the corresponding deoxyribonucleoside, deoxycytidine **11**.

No one would argue the importance of representing such aromatic/tautomeric systems, and being able to search for them. However, results of exchange of, and queries over, such structural types among different systems are difficult to predict.

Other bond properties. Different systems treat other types of bonding, for example, ionic, dative, multicentered, coordinate, etc., in different ways. Disagreements exist among chemists on how to draw such structures, let alone represent them in the computer. This makes it difficult, if not impossible, to interpret another system's handling of, for example, parent-salt forms, nitro groups, boron hydrides, ferrocenes, etc., unambiguously, without error. The structure of ferrocene provides an illustrative example. We have observed several different methods for drawing ferrocene, as shown in Figure 3.

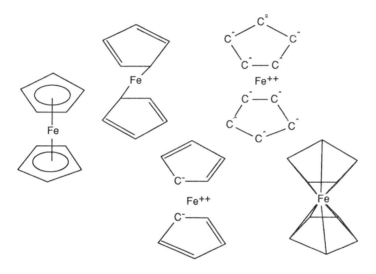

Figure 3. Alternative methods for drawing the structure of ferrocene.

Some representations are tidy drawings with no corresponding internal representation in any system. Others correspond to various degrees of compromise to allow storage at all, and may or may not be legitimate alternatives. Without detailed knowledge of the actual method chosen to represent ferrocene in a given system, it would be impossible to guarantee that one could find it (or verify its absence) in a data base.

Other atom properties. Intertwined with all the above problems are problems imposed by additional atom properties that must be represented. These include radicals, isotopes, and charges. Representing, or recognizing, the mobility of radicals and charges, interacting with resonance and tautomer forms, creates a real challenge *within* a single system; complete and correct exchange of this informatior between systems is problematic. Consider the problems posed by the coordination compound, Figure 4, composed of Co^{+3} together with three charged ligands which are capable of tautomerism. There are many ways to represent this structure graphically and internally, some of which are indicated in Figure 4.

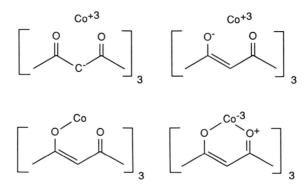

Figure 4. Some of the very many ways to draw a coordination compound with charged, tautomeric ligands.

The conclusion is quite clear. Without a very detailed specification of the representation and the subtle assumptions behind it, it is impossible to guarantee: (1) correct exchange of chemical structures among programs; and (2) correct retrieval in structure and substructure search, unless the query system matches the representation exactly.

Chemical Substances, Reactions, and Other Entities. Chemistry is not restricted to conventional organic structures. A large number of chemists and biochemists work with a wide variety of other chemical entities, and chemical reactions involving such entities. A partial list is provided in Table I.

Table I. Some classes of chemical substances for which computer representation may be important

Polymers	Mixtures
Coordination compounds	Salts
Pi complexes	Formulations
Metallic solids	Solutions
Alloys	Emulsions
Biopolymers	Colloids
Micelles	Crystalline solids
Membranes	Transition states

These substances, are treated incompletely, if at all, at the structural level by modern chemical information systems. Work is progressing rapidly in this area, including representational systems for conformations, polymers, mixtures, formulations, and so forth. Reactions of all these substances must be represented, another major challenge. Finally, there are other chemical entities which are important for certain computer-based applications, including Markush (generic) structures, and three-dimensional conformations of individual structures and substances. Space does not permit discussion of the many complexities involved in representing and searching such entities. Work on standards has only begun for the representing the simpler chemical structures and reactions. This work will inevitably lag far behind the rapid pace of software developments, leading, if we are not careful, to another round of incompatibilities in representation of chemical substances.

The Future – Bad News and Good News

The Bad News. There will remain strong forces that work against integration of information in general, and chemical information in particular. The most important of these forces are the following.

Standards. Standards will continue to emerge. The problem is that in all areas of information, from chemical structures to data to documents and text to images, there will be so many of them.

Hardware as a Commodity. The idea of compatible hardware platforms and the resulting emergence of hardware as a real commodity is a good one, but it will not be achieved soon if at all. Proprietary architectures will remain until computer manufacturers learn how to make money by cooperating rather than competing.

Operating Environments. Like standards, there will be several operating environments. The idiosyncrasies of each will make construction of machine and OE-independent user interfaces a very difficult task.

Application Software. Development of high quality software will continue to proceed at a pace slower than that of hardware and OEs. Software costs will continue to represent a larger fraction of an organization's computer budget.

Competition. The chemical community is itself a highly competitive one, whether one is engaged in academic research or in industrial production of new and better chemical substances. We should acknowledge that the spirit of competition exists as strongly in the production of high quality computer hardware, software, and data bases.

Leverage. The chemistry community is quite large, but it pales in comparison to other sectors of the economy for which hardware and OE vendors produce products. We have only limited leverage in getting those vendors to produce compatible systems that can be used as a common foundation for integrated application software.

Representation of Chemical Structures. Researchers in chemical information systems have spent many years attempting to get the continuous functions of electron densities in chemical bonds forced into the discrete representations demanded by digital computers. Problems will not go away as long as there are disagreements among chemists about fundamental aspects of representing chemistry in computers.

The Good News. Progress is being made in each of the problem areas outlined above. We can look forward to several advances that will make information integration much simpler in the future.

Standards. The emergence of some general standards where there are none now, coupled with pressures from consumers, will drive the adoption of standards that will make our lives easier. This has become the case for SQL-based RDBMSs. It will become the case for compound documents. Much remains to be done for text data base management systems, and image storage and retrieval systems. Standards will emerge to the extent that the community demands them. As they emerge, support them, we will all benefit!

Hardware as a Commodity. This problem will not be solved soon. The consumer will continue to be bombarded with choices of ever-increasing functionality and performance. Our hardest task is to distinguish between what may be possible with the new technologies, and what is actually practical, and realistic. Rather than doing nothing, we should recognize that the computing industry is moving rapidly to distributed systems, with networked servers and workstations. This implies two criteria for choosing hardware: (1) choose from vendors who offer powerful networking environments among their own hardware, and some well defined strategy for, and commitment to, smooth interfaces to other vendors' hardware; and (2) choose workstations with sufficient capacity to run the coming generation of operating environments. Any decision made may, in hindsight, be wrong, but you will obtain one to several years of use from such hardware.

Operating Environments. The problem of incompatible operating environments also will not be solved soon. However, vendors of application software will likely provide systems that run in similar ways, as far as the end-user is concerned, on different OEs. This is a real burden for the vendors, but will reduce the training costs for consumers substantially. Unfortunately, interfaces provided by different, competing vendors will probably remain different.

Application Software. Although the pace of development may increase only slowly, the functionality provided will increase dramatically. Every vendor will be required by consumers to provide an "open" architecture. Interfaces will be provided to allow a consumer to integrate various third-party software packages in new and different ways. Interfaces and overall system function will be customizable to suit the application and the needs of the end-user. Although this places more of a development and maintenance burden on the consumer, the advantages will greatly outweigh the disadvantages.

Quality of software will improve as consumers become less tolerant of "bugs." Software companies will devote an increasing portion of their budgets to quality assurance as the expectations of end-users increase. This may increase the cost of software, but will certainly decrease the overall costs to the consumers.

Competition. Competition will not go away. The community will suffer if it does, because the result may be lower quality software. In the absence of standards, especially in the chemistry community, systems will be produced that are incompatible. But it is probably better to get high quality software with a much higher level of functionality than to get complete compatibility with inadequate quality and performance. More compatible systems will be produced, because consumers want it. This expectation should, however, always be tempered by the realities of competition.

Leverage. The chemical community *is* large enough to get the attention of some vendors. This is certainly true for hardware, with DEC, IBM, HP, Apple and others targeting the community as an important vertical market. It is not true so far for vendors of OEs; they are first struggling to make the software work, and work efficiently. When they are done, they will look for important vertical markets. For application software, the community already has a relatively rich choice among several vendors. What is more important, however, is that several vendors of software for horizontal markets have defined our community as an important vertical market for their products. This is especially true for RDBMSs and software for document production, and will foster strategic relationships among vendors that will benefit the consumer.

Representation of Chemical Structures. We should not expect software for chemical information to solve the Schrödinger equation in order to derive complete descriptions of electron densities that accurately characterize chemical substances. In the absence of such detail, chemical representations among systems will inevitably have differences. This does not mean that efforts to produce a standard interchange format are misguided. All of the advantages and disadvantages of the analogy to compatibility among word processing systems, Figure 1, hold true for structure interchange. There are two efforts that have begun recently, one initiated in Europe, aimed at defining a Standard Molecular Data (SMD) format, the other begun in the United States, aimed at defining a Standard Format for Molecular Description Files. The groups involved are talking with one another. The problems they face include those outlined above, and are increased in scope by the U.S. group's greater emphasis on three-dimensional representations of structures.

There is an alternative approach, however, which is aimed at the technology for *searching* rather than at a common representation. In the foreseeable future, there will be many different methods chosen to represent chemical structures. The challenge then becomes to devise search techniques that allow a structure to be found independent of the representation chosen to store it in a data base. This approach would allow flexibility in representations used within a single system, or among several systems, as long as the representation chosen makes chemical sense.

Conclusion

Beginning many years ago, consumers, end-users, hardware manufacturers, and software developers, all made a large number of small decisions that collectively impede information exchange and integration today. We all know that now, and I hope that this paper has clarified some of the complex issues we face in achieving integration of chemical information in the future. Those consumers and developers who recognize and come to terms with these issues will be successful in the future.

Acknowledgments

I want to thank Jim Nourse, Doug Hounshell, Jim Dill and Jim Barstow at Molecular Design Ltd, who provided examples and valuable comments during the development of this paper.

RECEIVED May 7, 1989

Chapter 4

Chemical Structure Browsing

Alexander J. Lawson

Beilstein Institut, Varrentrappstr. 40-42, D-600 Frankfurt 90,
Federal Republic of Germany

Some of the aspects of structure browsing with the Lawson Number (LN) are described including limitations. Use of several LN in combination, single LN, and range-searching is demonstrated for the retrieval of various analogues, including positional isomers.

The purpose of this paper is to describe the implementation of a structure browsing tool in computerized data bases, with particular reference to the Beilstein data base as an example. The retrieval term which will form the main subject of this paper is the so-called "Lawson-Number" (LN). For obvious reasons, it is somewhat embarrassing for the present author to give an account of a descriptor which bears his own name. However, the driving force behind the choice of that particular name was not the author himself, but rather a combination of circumstances, the prime being that all "sensible" names for numbers of all descriptions had already been used up in the early days of the Beilstein Online venture. There were (and still are) the following terms in daily use in the production processes at the Beilstein Institute:

1. Beilstein Registry Number (BRN).
2. Identification Number.
3. CAS Registration Number.
4. System Number.
5. Concordance Number.
6. Formula Number.

The BRN is a number with no structural information whatsoever. It is assigned sequentially to each structure new to the registration software of the Beilstein Data Base. It is the primary key of Beilstein Online.

The Identification Number is a temporary (but unique) descriptor which accompanies each structure from the moment of its abstraction from the primary literature. It is specific for the combination of structure and citation. It is not searchable online, but is the single most important structural key for the production process at Beilstein.

The System Number is the unit of the Beilstein System of structure Classification. The values run from 1 to 4720. The System Number is printed on the page header of every odd page in the Beilstein Handbook, and (as a range) on the spines.

0097–6156/89/0400–0041$06.00/0

System Numbers are entirely dependent on structural features, but are nonunique. System Numbers are not searchable online; their function has been taken over by the LN, but the numeric value of the System number does not correspond to that of the LN.

The concordance numbers are the cross referencing page numbers used to indicate the (theoretical or actual) pages on which entries on any particular structure would have been published in earlier Supplementary Series of the Beilstein Handbook. Concordance numbers are entirely dependent on structural features. Concordance numbers are searchable online (in the form of the Source Code SO).

The formula number is the key to the printed structure graphics of the Beilstein Handbook. It acts as a link between text and graphics at all stages in the Handbook production. It is not searchable online.

The LN is a recast version of the System Number, made to fit the computer age. Perhaps it would have been better to create the term "New System Number", but since the number stems indirectly from the author's SANDRA (1) algorithm, the term "Lawson-Number" came into being on the suggestion of the Beilstein Online President C. Jochum, and such terms develop a life of their own.

The simplest definition of the idea behind the LN is as follows: to provide a retrieval tool for computerized chemical data bases which would allow a degree of structure browsing comparable to that long enjoyed in the Beilstein Handbook. All who have searched the Beilstein Handbook know what is meant by this browsing, and how powerful this can be. The reader is referred to the excellent article (2) by David Bawden, "Information Systems and the Stimulation of Creativity", which contains an excellent description of browsing, and to the outstanding contributions (3) of Peter Willett in this field. Bawden states that "the little-understood browsing function is the single most important means of creative use of the literature, whether in printed or computerized form".

If the LN can contribute to the stimulation of creativity, then the present author will feel satisfied. However, browsing in computerized data bases is a dangerous field; two of the most obvious pitfalls are as follows.

Browsing implies similarity and "fuzzy" data, and both of these terms involve a subjective appraisal of the "in-context" value of the data retrieved. What is similar to one researcher will be false drops to another, or even to the same researcher in a different context. Browsing therefore always implies subsequent screening (in an online context, by the use of further search terms). Similarly, browsing can never be exhaustive in all dimensions of similarity at one and the same time.

The online researcher generally works with a very specific goal, and is very often working as intermediary to the true "end-user". In any event, he is accustomed to being offered a precise definition of the search tool which he is about to use (and pay for). He therefore develops a search strategy before using the medium. This, incidentally, is not necessarily the case for the user of printed works. Here we encounter a second difficulty. Browsing (by definition) must be able to generate hits which fall outside the expectations of the searcher, i.e., outside of his strategy. Therefore, serendipity is not a natural consequence of online use, and is perhaps even incompatible with the medium. The present author would prefer to hope that the latter is not true, but recognizes the difficulties involved.

There is no effective answer to the first of these problems, other than the obvious one, that the human eye selects the significant from the insignificant in a quite remarkable way, provided that sufficient prescreening is achieved.

In the case of the second problem, however, it is both possible and desirable to give a definition of the search tool, and that is the purpose of this paper. The LN is a flexible and useful tool, but contains the seeds of much difficulty if the concept is applied incautiously. It is not a type of substructure searching, nor is it a Markush search. To repeat the above objective once more, the intention behind the LN is to retrieve those sorts of structures which one might have found within a few pages of the given structure, had the user been looking in an Omnibus Edition of the Beilstein Handbook.

Defined in this way, the LN becomes more readily understandable. The sort of compounds always found in the vicinity of any given compound in the Handbook are as follows (in order of increasing browsing distance):

1. Stereoisomers.
2. Functional derivatives (esters, ethers).
3. Positional isomers.
4. Halogenated and/or nitrated derivatives.
5. Other analogues.

The LN should handle these groups of structural relationships, but also be coupled with a "feeling" for the amount of data which the eye can efficiently browse. The following sections explain how this problem was tackled, and show some examples of use. It is convenient to discuss the design and limitations of the LN before proceeding to an example.

Design of the LN

The first problem to be tackled in the design of the LN was the question of statistical distribution of hits. Ideally, any LN should return approximately the same number of hits from a data base as any second LN. Furthermore, the range of LN used should be finite and fixed, and yet contain values for any conceivable structure. This problem was solved by the use of the Beilstein Handbook pages as a normalizing factor, as described elsewhere (*1, 4*). In effect, the whole field of possible structure fragments in organic chemistry was divided into a fixed field of 32768 regions, and each region received a number.

The LN is therefore a fragment code, with values (at present) between 0 and 32767. In practice, only the values between 9 and 32759 are used; any given LN should retrieve approximately 1/20000 part of the total data base, with certain unavoidable exceptions (e.g., methoxy, see below).

Each fully defined organic structure is characterized in a selective (but non-unique) manner by a number of LN, usually of the order of 1 to 5, typically 2 to 3.

For instance: LN = 26594; 2826 (Compound 1 of Figure 2) where 26594 is the heterocyclic unit and its dicyano side chain (recognized as a masked diacid) and 2826 is the Et-N fragment

Each LN refers to a separate fragment in the molecule, and any particular fragment will always have the same value of the LN.

However, it is generally not true that any one LN refers to one fragment alone.

On closer inspection, the members of the group possessing any one value of the LN will be seen to fall into distinct subgroups, or families (stereoisomers, positional isomers, analogues etc., as discussed above).

This can be seen best in schematic form in Figure 1.

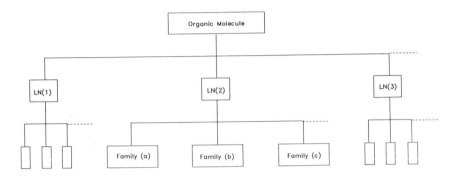

Figure 1. LN as a common link.

Thus it can be seen that the LN is a common link between any particular molecule and a certain number of families of molecules, each possessing certain combinations of structural aspects in common. The nature of these structural aspects is listed below in Table I, in the order of their influence on the value of the LN.

For example, the "cyclic class" aspect dictates that:

1. All acyclic fragments have LN 9 → 3815.
2. All carbocyclic fragments have LN 3816 → 16767.
3. All heterocyclic fragments have LN 16768 → 32759.

Table I: Factors Governing the Value of the LN

1. Cyclic class (including number and type of heteroatoms).
2. Chemical functions (amine, hydroxy . . .).
3. Degree of unsaturation of the carbon framework, measured in terms of the multiple bonds at carbon + ring closures.
4. Carbon-count of the carbon-complete fragment framework.
5. Degree of carbon-branching (butyl, sec-butyl, tert-butyl . . .).
6. Degree of halogen and nitro substitution.
7. Chalcogen exchange (oxygen replaced by sulfur, selenium etc.).
8. Ring sizes (azulene, naphthalene . . .).

Limitations of the LN

The numerical value of the LN is dependent *only* on the presence of the above structural features, according to a set of clearly defined rules. These rules are based on the Beilstein System, which is a thoroughly tested algorithm for the classification of organic structures. However, a knowledge of the System is not required for the use of the LN. Nevertheless, for the purpose of this introduction to the LN it is helpful to understand certain aspects of these rules, so that the limitations of the approach can be properly appreciated.

Four limiting aspects are especially important:

Carbon-Completeness. The rules demand that any given molecule should be treated as an aggregate of one or more *carbon*-based fragments. Let us call these fragments Beilstein Registry Fragments (BRF) for the moment.

The rule is that each BRF extends over all carbon-to-carbon bonds, and cannot be subdivided by breaking a carbon-to-carbon connectivity. The boundary points for a BRF are exocyclic heteroatoms. (For a more detailed discussion, see reference 4 and references therein).

For instance, anisole (Ph−O−Me) is composed of only two BRF, namely the fragment for Ph−O and the fragment O−Me (note that the linking heteroatom oxygen atom is assigned to *both* BRF).

Thus the LN values for anisole would be 5219 and 289, where the LN for Ph−O is 5219, and the LN for Me−O is 289.

It follows that the value of an LN can only contain information about the carbon-complete fragment: a query with the LN value of allyl alcohol (C=C−C−O) cannot give anisole as a hit. The LN is not intended to be a substructure tool (other systems are available for this, obviously).

Combined Functionality. The system lays great emphasis on functionality, both in terms of skeletal morphology (e.g., ring features) and functional groups. However, functionality is represented as the sum of the separate parts; e.g., the C6−carbocyclic nature of the Ph−O group, the unsaturation (3 formal double bonds) and its hydroxy function are combined in the LN value. This means that these are inseparably mixed together in the number 5219 and cannot be separately addressed. The LN is not only carbon-complete, it is also "function-complete" for the BRF.

The LN is not Immediately Transparent. All the above 8 factors find expression in a single number, and it would be impossible to make this number react in transparent manner as a function of changing any combination of the factors. In other words, there is no obvious relation between the LN value for Ph−O and that for hydroxy-biphenyl (Ph−Ph−O), since a number of factors are different (degree of unsaturation, carbon count . . .). On the other hand, the LNs of any given fragment can be generated in a few seconds by a simple PC computer program, direct from the drawn structures; this is ideal for those who wish to be able to generate the LNs for any given molecule, but the purpose of this article is to demonstrate that the LN field may also be used without this knowledge, to considerable effect.

It is not a Unique Number for Unique Fragments. This point should be stressed, even at the risk of becoming repetitive. The individual LN are *not* unique values for unique fragments (since there are clearly more than 32768 organic fragments possible), but are reproducible quantities for any given fragment. Thus the fragment 26594 (for instance) will define a potentially infinite group of fragments, but the members of this group will have certain distinct structural characteristics in common. This actual example is shown below to illustrate the various points made in this paper.

Example of the Use of the LN

Use of Single LN. The LN value 26594 was used as a search term: the 16 hits of Figures 2 and 3 were retrieved from a data base corresponding to approximately 1 million structures (and are reproduced by kind permission of The Scientific & Technical Network, STN), along with ten further hits (not shown), which were

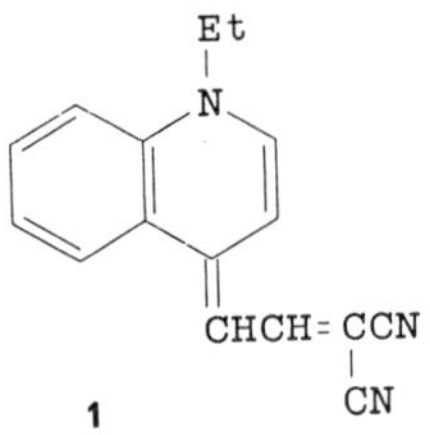

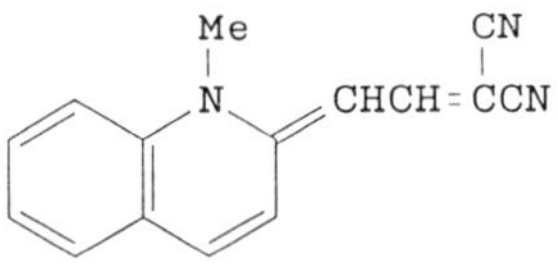

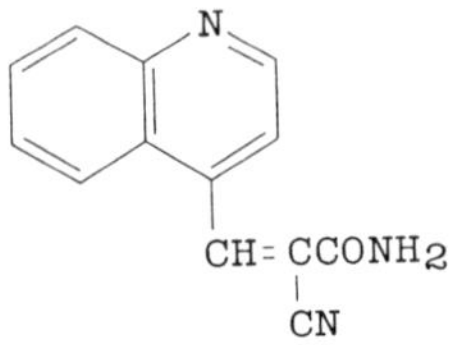

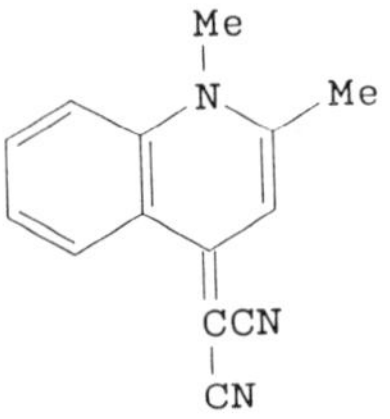

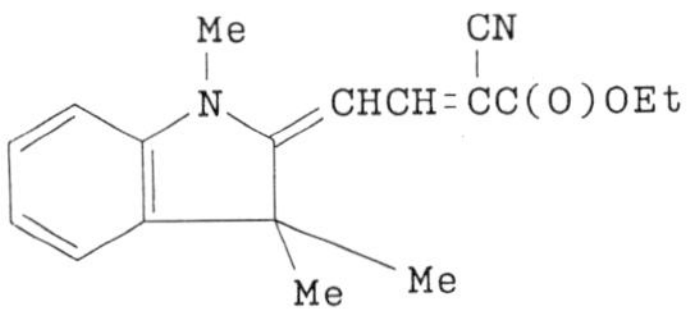

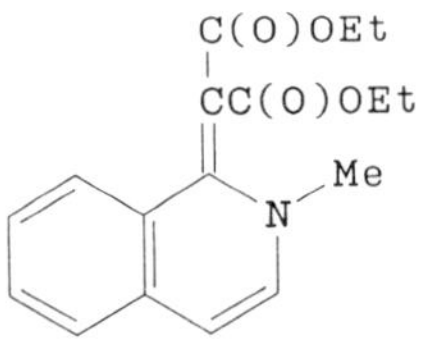

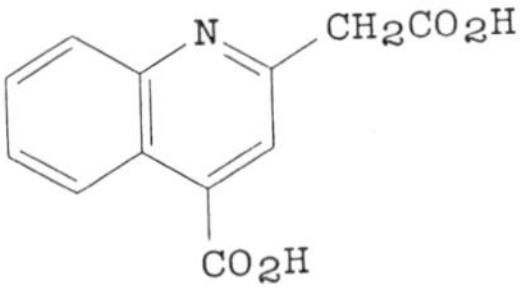

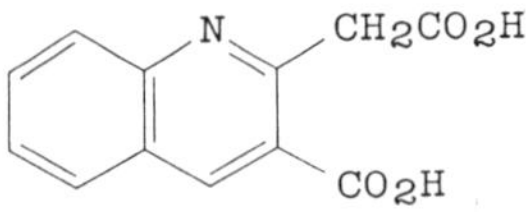

Figure 2. Bicyclic compounds with LN 26594.

mainly esters and amides of the acids, along with a few further analogues. Thus the resolution of this LN is fairly typical, and in fact is better than the expected 1/20000.

Visual examination of the hits of Figure 2 (bicyclic) and Figure 3 (tricyclic) illustrates the browsing effect more clearly than a simple listing of the various features. In particular, it should be clear to the reader how different this result is compared to a substructure search. There is a constant background of mononitrogen heterocycles containing two carboxylic groups, either in free form or as esters, amides, even nitriles. Although no single conventional description defines this group of similar compounds, there are clearly several groups which can be further separated by the use of nomenclature search terms (quinolines, indoles . . .). In general it is useful to use the LN in logical combination with another search term

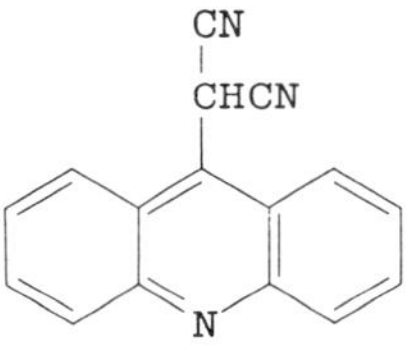

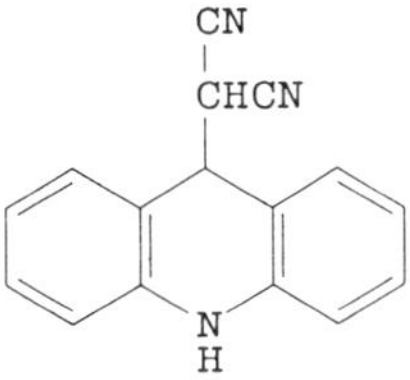

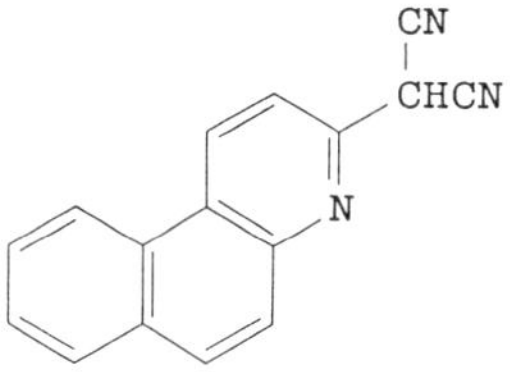

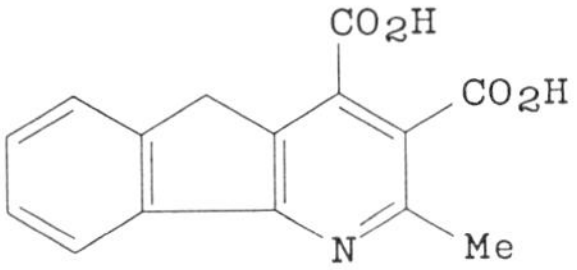

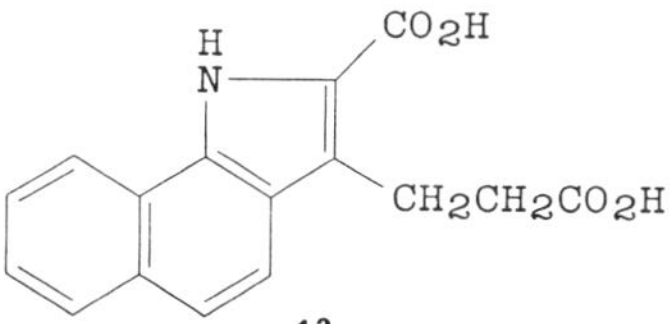

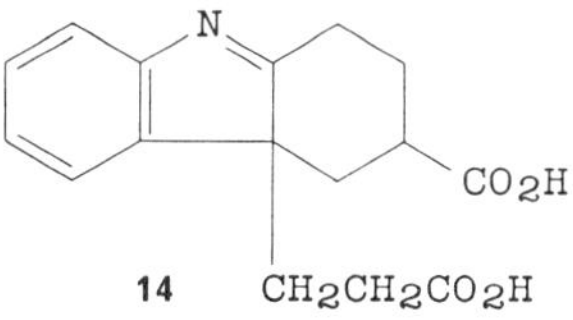

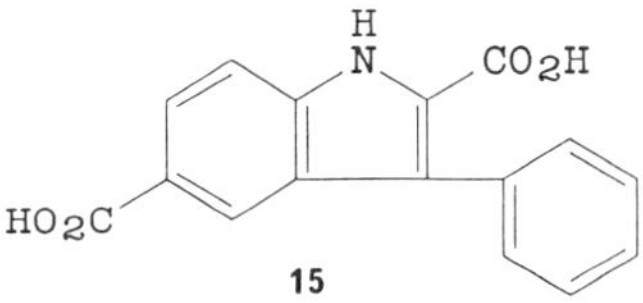

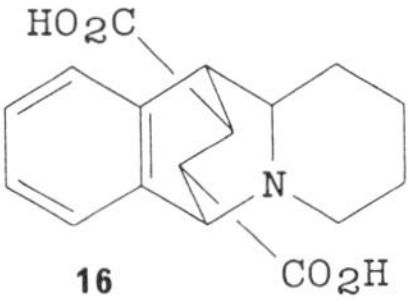

Figure 3. Tricyclic compounds with LN 26594.

(e.g., nomenclature terms, or a term from the molecular formula, or even a second LN). The next two subsections below illustrate this point.

The use of a single LN (as here) always guarantees the following to be among the hits (if present in the data base):

1. Stereoisomers.
2. From acids: esters, amides, acid chlorides.
3. From hydroxy and mercapto fragments: ethers.
4. From amines: primary to quaternary analogues.
5. Positional isomers.

Thus stereoisomers of compound **14**, the free acid (and monoester) of compound **6** would also have the value 26594. Note that the LN is a reliable retrieval tool for positional isomers, such as compounds **7** and **8** (also **1** and **2** in the heterocyclic BRF). Furthermore, it is legitimate to state that there are no further positional isomers of these particular species in the data base, otherwise they would have been hits in this list. Similarly, the data base contains no further examples of X-phenyl-indole-Y,Z-dicarboxylic acid (compound **15**).

Use of Several LN in Combination. Let us go back again briefly to the example of anisole, mentioned above. In the following we shall use the query language of the Scientific and Technical Network (STN). The query:

s 289/LN

would give every compound in the data base with a methoxy subunit. This is an example of a very specific fragment query, which happens to have an extremely unselective effect, since approximately 14% of all known compounds contain the methoxy fragment, either in the form of methyl ethers or methyl esters. Similarly, the fragment Ph$-$O occurs in a vast number of compounds. But the combined query

s 5219/LN and 289/LN

would give all compounds of the general form Ph$-$O$-$ . . . $-$O$-$Me (several hundred). A truly selective query for anisole would be:

s 5219/LN and 289/LN and C=7

Obviously this particular example is neither typical nor very useful in practice, but serves to show the principle.

More simply, the combined use of the LN for Me$-$N (2817)

s 26594/LN AND 2817/LN

would have restricted the hits of Figures 2 and 3 to compounds **2,4,5** and **6**.

Range-Searching with the LN. All nitrated and C-halogenated derivatives of any given structure can be found in the immediate vicinity of the LN, in the so-called

"Range of 8". This involves use of the LN in a range-searching mode. The range of 8 is defined by the following procedure:

1. Divide the last 3 digits of the given LN by 8.
2. Substract the remainder from the given LN to define the start of the range.
3. Add 7 to this number to define end of range.

For example, for LN=26594 the range of 8 is found as follows:

$$594/8 = \text{remainder } 2$$
$$\text{range of } 8 = 26592 \text{ to } 26599.$$

Then the query:

$$\text{s } 26592\text{--}26599/\text{LN AND x/ELS and C=12}$$

will contain (among other analogues) all halogenated derivatives of all positional isomers of compounds **7** and **8** (if present), but not the esters. This arises from the fact that the terms ELS (element symbol) and C (number of carbons in the molecular formula) refer to the complete molecule, while the LN refers to the BRF (a subset of the molecule).

In actual fact this query results in two dihalogen derivatives of 2-methyl-quinoline-3,4-dicarboxylic acid. It is therefore clear that C-halogenated derivatives of free X-carboxymethyl-quinoline-Y-carboxylic acids are not present in the data base. This search technique is independent of the vagaries of chemical nomenclature: the use of alternative names based on X-carboxy-quinol-Y-yl-acetic acids (etc.) is avoided.

Conclusion

The LN is one more tool available for the online searcher. It is intended to be complementary to substructure searching. Used in judicious combination with other search terms it can simulate the browsing effect normally associated with printed works.

Acknowledgments

The author thanks the West German Bundesministerium für Forschung and Technologie for financial support.

Literature Cited

1. Lawson, A.J. In *Graphics for Chemical Structures*; Warr, W.A., Ed.; ACS Symposium Series 341; American Chemical Society: Washington, D.C., 1987, pp 80–87.
2. Bawden, D. *J. Inf. Sci.* **1986**, *12*, 203–216.
3. Willett, P. *Similarity and Clustering Methods in Chemical Information Systems*; Research Studies Press: Letchworth, 1987.
4. Lawson, A.J. In *Software-Entwicklung in der Chemie 2*; Gasteiger, J., Ed.; Springer Verlag: Heidelberg, 1988, p 1.

RECEIVED May 2, 1989

Chapter 5

Integration and Standards: Use of a Host Language Interface

**A. Peter Johnson, Katherine Burt, Anthony P. F. Cook,
Kevin M. Higgins, Glen A. Hopkinson and Gurmaj Singh**
ORAC Ltd., 175 Woodhouse Lane, Leeds LS2 3AR, England

ORAC (Organic Reactions Accessed by Computer) and OSAC
(Organic Structures Accessed by Computer) have developed into
mature systems for the handling of reaction data and structural data.
The next phase of their development is concerned with the seamless
integration of these systems with other computer-based laboratory
tools. We have made considerable progress in this area. The Host
Language Interface allows other programs to interrogate the ORAC/
OSAC data bases in order to retrieve data which can then be
manipulated appropriately. Currently, this provides a user-friendly
method for the integration of relational data base management systems
with the OSAC system. The IMPORT and EXPORT facilities allow
data to be transferred to or from these systems in the Standard
Molecular Data (SMD) format. Currently the import facility is used to
upload reaction data which has been created on a PC in SMD format,
using the program PsiORAC. Further work in this area will be
concerned with extension of the SMD format to deal with queries, and
the use of this extended SMD within the Host Language Interface to
provide open interconnections to a variety of systems, including those
concerned with synthesis planning, (LHASA and CASP), product
prediction (CAMEO), quantitative structure activity relationships
(QSAR) and modeling.

The goal of a fully integrated chemical and biological information system is one that
is being vigorously pursued in a number of industrial laboratories. While there are
bound to be some differences of opinion as to precisely which components should
form an essential part of such a system, Figure 1 depicts some of the modules which
might be of value to a large pharmaceutical research laboratory. A reaction data
base management system (such as ORAC, REACCS or SYNLIB) and a chemical
structure data base management system (such as OSAC, MACCS, DARC-SMS or
HTSS) might be at the heart of such a system but the latter would have strong links
to a general purpose data base management system (such as ORACLE, RDB,
System 1032 or Ingres) which would typically handle biological and toxicological
data. The reaction data base system could be usefully linked to a synthesis planning

0097–6156/89/0400–0050$06.00/0

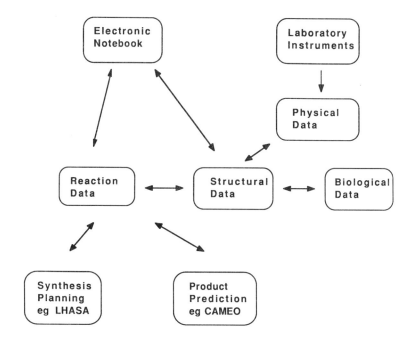

Figure 1. System modules

program, such as LHASA (*1*), CASP (*2*) or EROS (*3*) and also to a reaction prediction system such as CAMEO (*4*). In each case the reaction data base could be used to retrieve automatically literature examples of the reactions suggested by these other systems. Finally, the module labeled "electronic notebook" could take a variety of forms, but perhaps the various suites of PC programs (e.g., Molecular Design Limited's CPSS or Hampden Data Services's PSIDOM) provide the closest approximation that we have at present to an electronic notebook.

The term "integration" used in relation to these various systems could involve a number of possible scenarios, but should at the very minimum include:

1. The ability to transfer smoothly data from one system to another.
2. The ability to use a single complex query, the elements of which require the search of both a structure data base and a general purpose data base, for example.
3. The ability to display on a screen (or report) data which has been derived from two or more different systems.

In the ORAC/OSAC context the achievement of these goals has been complicated by the flexibility built into both systems. Thus both systems feature:

1. User-definable query menus.
2. User-definable display formats.

3. Range of available data types e.g., integer, real, character (free text or controlled thesaurus).
4. User-definable hierarchical thesauri.

Notwithstanding these problems, these goals have been achieved in the ORAC/OSAC context.

Data Transfer

Data transfer between software systems created by a single vendor is usually fairly straightforward. Where more than one vendor is involved, the absence of any agreed standard can make the development of software to accomplish such transfers a needlessly time-consuming process. For this reason we strongly support the development of an interchange standard and indeed have recently developed software utilities which permit the transfer of data either from an ORAC/OSAC data base to a Standard Molecular Data (SMD) file (SMD EXPORT) or from an SMD file to an ORAC/OSAC data base (SMD IMPORT).

Two examples will serve to illustrate the value of the SMD IMPORT utility. The Hampden Data Services PsiORAC software permits a complete reaction data card to be treated on a PC in SMD format. We have now used SMD IMPORT to load some 10,000 such cards into an ORAC data base (held on a DEC VAX computer). The second example involves the automatic generation of an ORAC database of 40,000 reactions from raw data contained in (1) a file of starting material and product structures held in a DARC-SMS data base, and (2) an ASCII file containing pointers to these structures plus additional data. With a modest amount of effort, a conversion program was written which, as shown in Figure 2, made use of an existing DARC to SMD utility and also merged the structures with the other data to create a reaction data base in SMD format. This was then loaded into an ORAC data base in the usual way.

Our work on these problems has enabled us to identify some of the drawbacks of the current SMD format and we will certainly work with other interested parties to try to improve this standard.

Integration at the Search and Display Level

The key component in our solution to the problem of integration is the ODAC Host Language Interface (HLI). ODAC (Organic Data Accessed by Computer) is a generic term which includes ORAC and OSAC. The Host Language Interface is a module which permits a range of different external application programs to have access to all of the functionality embedded in the ORAC and OSAC programs. Some of the functions which can be accessed by HLI are shown in Figure 3. The requirement to OPEN or CLOSE one or more data bases follows on from the ODAC facility which permits concurrent search of a number of data bases. HLI can output structural and other data to the controlling applications program, but control of the final format for display or plotting resides in the applications program so that display or plot formats can be easily customized. This is particularly useful

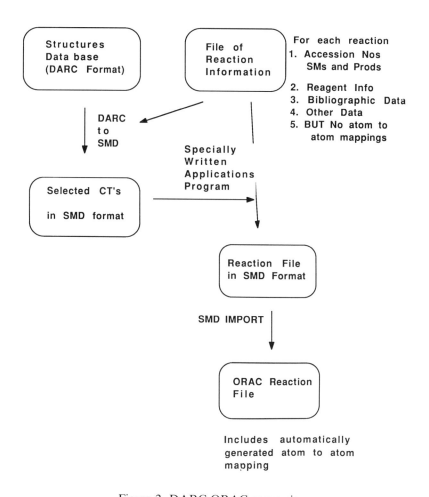

Figure 2. DARC-ORAC conversion

when displaying information coming from two or more different data base management systems.

Figures 4 and 5 indicate schematically the organization of the search and reporting facilities respectively.

As developed, HLI is a general purpose interface between ODAC and any external program. The number of uses to which such an interface can be put is limited only by one's imagination. Two recent examples of the use that we and our collaborators have made of HLI are described here.

The first example concerns the addition of text concerning experimental procedures to our now complete ORAC version of the Theilheimer reaction data base. The initially released version lacked this additional information and it was felt

Begin/End OSAC/ORAC;

Open/Close databoxes;

Store/Retrieve/Modify   data;

Define complex search queries;

Perform data base searches;

Manipulate answer sets;

Output data and structures.

Figure 3. HLI main functions

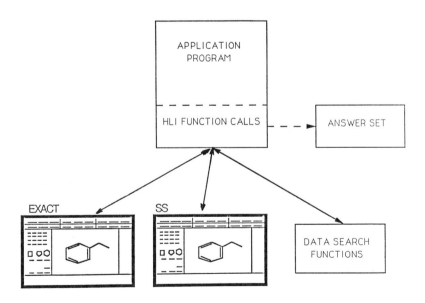

Figure 4. Data base searching using the HLI

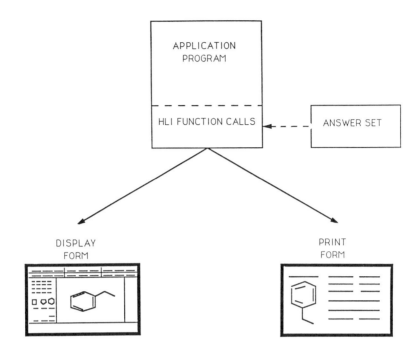

Figure 5. Data and structure output using the HLI

that its inclusion would enhance the utility of the data base. One way in which this could be accomplished is to use the UPDATE facility in ORAC to add the data interactively to each reaction card. In practice, it has proved to be much more efficient to create an ASCII file containing the additional data for a large number of reactions, and then use an application program (written in just a few days) to load the data into the ORAC data base *via* HLI. The overall process is shown in Figure 6.

The second example is altogether more substantial and concerns the use of HLI to integrate structure search using OSAC with biological data search using SYSTEM 1032. The system (5) developed in collaboration with one of the major OSAC industrial users is depicted in Figure 7. ABACUS (Advanced Biological and Chemical Unified System) is the application program, written in 1032 code, which provides a user interface offering apparently seamless integration of search and display of data from either or both of the underlying search systems. Not surprisingly, HLI provides the gateway into OSAC.

Integration with Synthesis Planning Systems

The wide range of functionality offered by HLI should make it the ideal vehicle for interfacing synthesis planning systems such as LHASA to reaction and structure data bases. Some obvious applications are shown in Figure 8 and include fetching and displaying real examples drawn from the literature of the reactions suggested by LHASA. The mechanics of this process would involve automatic generation of a reaction query within LHASA and then passing this query to OSAC *via* HLI. This is not possible at present because we do not yet have a language in which that structure-based reaction query could be transmitted from LHASA to ORAC. What is needed is an external format for structure query representation, one of the topics which the SMD user group is trying to address. It would be very easy for us to devise our own unique language for structure queries, but we would clearly prefer to use a more widely accepted convention.

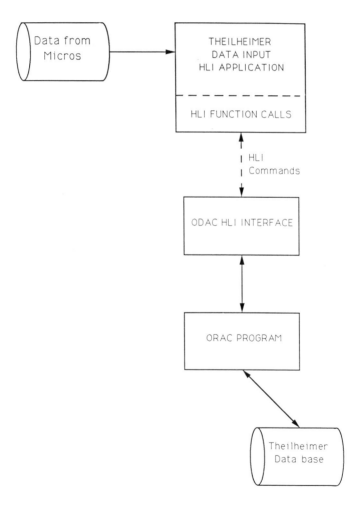

Figure 6. Adding Theilheimer data to ORAC using the HLI

INTEGRATED CHEMICAL
INFORMATION SYSTEMS

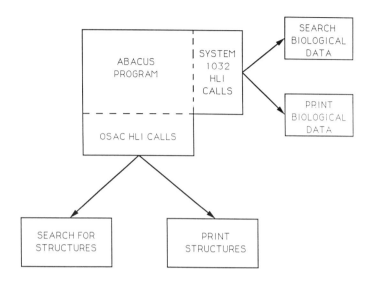

Figure 7. Potential application of the ODAC HLI

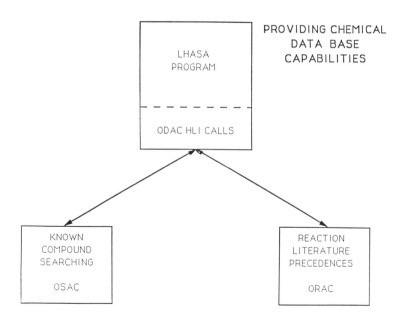

Figure 8. Potential application of the ODAC HLI

Conclusion

In conclusion, as indicated in Figure 9, the HLI concept is the cornerstone of our integration plans. It provides a consistent, stable and easily maintainable interface for both OSAC and ORAC which has already proved its value, and we are confident that it will continue to be a key component in our development plans for some considerable time.

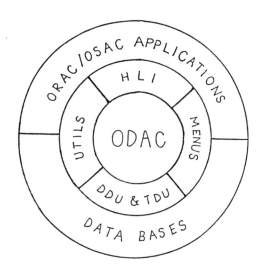

Figure 9. The HLI concept in integration

Literature Cited

1. Corey, E.J., Wipke, W.T. *Science* **1969**, *166*, 178–192.
2. Sieber, W. In *Chemical Structures: the International Language of Chemistry*; Warr, W.A., Ed.; Springer Verlag: Heidelberg, 1988; pp 361–366.
3. Gasteiger, J.; Hutchings, M.G.; Saller, H.; Löw, P. In *Chemical Structures: The International Language of Chemistry*; Warr, W.A., Ed.; Springer Verlag: Heidelberg, 1988; pp 343–359.
4. Gushurst, A.J.; Jorgensen, W.L. *J. Org. Chem.* **1988**, *53*, 3397–3408.
5. Magrill, D.S. In *Chemical Structures: The International Language of Chemistry*; Warr, W.A., Ed.; Springer Verlag: Heidelberg, 1988; pp 53–62.

RECEIVED May 2, 1989

Chapter 6

A Standard Interface to Public, Corporate, and Personal Files

John L. Macko and James V. Seals

Chemical Abstracts Service, 2640 Olentangy River Road, Columbus, OH 43210

Today's users of scientific and technical information have access to a wide choice of public, corporate, and personal information systems. Access is gained through myriad configurations of hardware, software and networks. Interface configurations that work for one system often do not work for another. This situation presents a formidable barrier to those who wish to retrieve information online, but do not wish to become information specialists. It largely explains the reluctance of end-users to do their own searching, and the reluctance of their companies to pay for their training. The long-term solution lies in a standard interface to public, corporate and personal systems that is easy to learn and easy to use. The interface software must facilitate interaction with files of bibliographic information, full text, numeric data, chemical structure information, and graphic images. It must be inexpensive and must run on a wide variety of computers. It must be designed to encourage offline query formulation and downloading of information for offline review and manipulation. It must be supported by established online services in America, Asia, and Europe, who are committed to that interface as a long-term international standard.

In referring to a "standard interface", we mean a software package that is widely used by scientists and engineers at their workstations to access public, corporate, and personal data bases. It is one kind of software in the growing array of software products that serve many special needs. We will explore briefly with you the concept of standards, the bridges that must be built before a standard interface can be created, some of the features we believe necessary for a standard interface, and some of the special difficulties involved in creating it.

Let us first consider why standards are necessary. Those who travel internationally notice immediately what a nuisance the lack of standards can be. For example, in the United States electrical appliances are made for outlets of 110 to 120 volts and 60 hertz. In Europe, they are made for 220 to 240 volts and 50 hertz. But even within Europe there is no standard for so simple a matter as the shape of the plug, so you may find that your appliance works in one country but not in the next. Similarly, you will have noticed that word processing hardware and software is not standardized even within a monolingual nation like the United States. A person stopped by the CAS office at the Los Angeles ACS meeting and asked to use our word processing equipment, but he quickly found that it was quite different from what he was accustomed to using. As a result, the keyboard and various features of the system just did not work for him. So, despite being well versed in word processing, he simply could not use our equipment without further training.

Standards exist everywhere in our daily life, but they are not obvious except where they are lacking. Without the existence of standard time, a standard alphabet, and standard units of measurement, life would be extremely difficult. When they are set properly, standards facilitate communication and interaction; they eliminate the need for continual retraining and give the consumer more flexibility. However, standards can also reduce flexibility, and if we do not adopt them carefully, they can stifle innovation or quickly become obsolete.

Standards for scientific software must be selected with particular care, because this area is growing rapidly and will change even more rapidly in the future. The standards we adopt must be solid, robust, and flexible enough to serve for as long as possible and flow with the advances in technology. In the chemical information field, the exchange format for chemical structures is particularly important and at present this function is not standardized. At CAS we believe the Standardized Molecular Data (SMD) movement appears, at the moment, to be a most promising development that could lead to a standard exchange format for chemical structures. CAS has not adopted SMD, but I am pleased to announce that CAS will support the program, with the objective of arriving at an exchange format that the user community could adopt as a standard for the exchange of chemical structures.

Notice that the title of this presentation refers not to "the" standard interface but to "a" standard interface. Many software packages have been introduced, and more are appearing continually, but no standard has yet been accepted. For a standard software for public, corporate, and personal files to be accepted, certain bridges must have been built. These will require not only cooperation but also commitment, and the relationship between the online data base vendors and front-end software developers will be crucial for success.

We online vendors cannot poll all those who have written front-end software packages to determine how changes we are planning might affect the front-end software they wrote. Even if this were possible, the online vendors could not promise to accommodate the front-end developers in every case. Accordingly, an independent organization that introduces software which they claim to be compatible with STN International, for example, might be making an accurate claim today. But even an insignificant change in STN's user interaction protocol could inadvertently make the front-end software suddenly unusable. Software developers who claim their product is compatible with all online vendors are especially vulnerable to the effects of such necessary changes. STN and other online vendors cannot always take the requirements of these independent software developers into account, so neither we nor they can guarantee the continued compatibility of their software with our systems.

Bridges must be built: only a bridge between the online vendor and the front-end software developer can assure long-term compatibility. Bridge in this case means a commitment, by both parties, to change only in ways that maintain compatibility between vendor software and the front-end software. Of course, this commitment already exists between STN International and its own STN Express software package. Another example of such a bridge in the making is the strategic alliance between CAS and Evans and Sutherland: as announced at the ACS national meeting last June, CAS and Evans and Sutherland will work toward the joint development of software that links STN Express with the molecular modeling and other capabilities of Evans and Sutherland software. We are uniting Alchemy and Concord products with STN Express so that, soon, users will have the capability to view structures from the CAS Registry File in three dimensions, and translate three-dimensional structures into two-dimensional representation for searching in the CAS Registry File. CAS and Evans and Sutherland are committed to maintain long-term compatibility.

Now let us consider in more detail the features a standard interface should have, the special difficulties it presents, and the kind of bridges that must be built.

Necessary Features

Compatibility. Ideally, a standard interface software package must run at the user's site on the computer of his or her preference. This may be a personal computer or a mid-range computer, but in any case compatibility is a key feature. By compatibility we mean the ability to interface with public or corporate online systems, a variety of networks, various brands of hardware, operating systems and applications software, and with different kinds of users who need many types of data.

Networks alone present a major challenge. The standard software must interact with both Local Area Networks (LANs) and Value Added Networks (VANs), operate at different line speeds, and handle different protocols. STN Express, for example, communicates not only with networks familiar in the United States, such as Compuserve, TELENET, and TYMNET, but also with Austpac, Datapak, DATEX P, IPSS, ITAPAK, TELEPAC, and TRANSPAC.

Hardware environments are another challenge to compatibility, especially since the most obvious choice can be risky. When we developed STN Express it was obvious that this software should be compatible with the IBM PC. But since then, the Macintosh has become increasingly popular. There are fads and fashions in the computer market just as in other areas. Today, front-end software should work not only with the IBM and Apple PCs but also with the DEC, Commodore, and NEC, to name a few. Preferences vary over time and also by country. We have seen that the Commodore is rather popular in Germany and the NEC seems to dominate the Japanese market. Those brands are virtually unknown in the United States, while the IBM PC is relatively unfamiliar in Japan. STN Express currently works with MS-DOS; the standard interface should also work with, at least, the UNIX and Macintosh operating systems. Perhaps more will be necessary in the future.

Applications software includes packages for preprocessing; for example, word processing and communications software that comes into play before the user retrieves data from an online data base. Users may wish to combine the use of these software functions in connection with their online searching. In addition, there is postprocessing software for handling the data in various ways after it is retrieved and downloaded. A user's ultimate goal in searching may be to produce a report, integrating information from different sources. Doing this may involve eliminating duplicate answers, sorting answer sets, combining results from different searches, and integrating this information with word processing and spreadsheet software, as well as packages for statistical analysis or molecular modeling. The standard interface must work with all of these and others.

Another important compatibility, of course, is with the search and retrieval software of the online system. This can mean both a public and a corporate system serving the same user. Similarly, the user may wish to interface with computation support software, which may be operated on both public and corporate systems.

The kind of data retrieved is another important consideration. Control characters are different in different systems, so this must be taken into account. The kinds of data a scientist might search and retrieve include not only text but also images, chemical structures, and numerics. In the United States, we use Latin characters generally, but chemistry and other fields require Greek and other special characters as well. But even handling text alone is not as simple as it may sound: remember, the English-speaking world is only part of the audience. We believe a truly universal interface eventually should be able to handle the Japanese, Chinese, and Cyrillic alphabets too.

Chemical structures are important data that are handled quite differently by various online systems. The variability of the connection table formats is a crucial difference in this regard, so the standard interface must handle whatever format is specific to a given online system. Some standardization and interchangeability in regard to connection tables will be necessary.

Numeric data have received too little attention in the past, as compared to bibliographic data bases. The organizations that operate STN International, JICST (the Japan Information Center of Science and Technology), FIZ Karlsruhe, and CAS, have undertaken a major project to develop numeric handling capabilities, and considerable progress has already been made. It must be remembered that handling numbers is quite different from handling text, and numeric data searching is more complicated than simply retrieving numbers given within text. Different operators are required, including "greater than," "less than," and "equal to." And the user might wish the software to interface with statistical packages, packages that plot graphs from numerical data, and many other software options.

Images, the search and retrieval expression of "pictures," present a challenge because of the sheer number of bits involved in producing these. Sending pictures across VANs is very expensive and very slow, even though the data are compacted. But if compacted, the data must then be uncompacted at the user's site, and this requires another level of compatibility between the front-end software and the applications software of the public or corporate search system.

User compatibility is at least as important as hardware and software compatibility. In an international community, users differ according to computer expertise, native tongue, and subject speciality, to name just three crucial aspects. Failure to account for any of these can make the front-end software less than fully functional for a given user.

The software should be designed for users with different levels of expertise in computer searching. It should be noted that "computer expertise" and "searching expertise" actually mean two different things. A generally computer-literate person may nevertheless be a semi-literate searcher. A sophisticated user should not be insulted by oversimplified options that do not permit taking full advantage of the online search system. On the other hand, the completely uninitiated user should be guided by the software in such a way that he or she will find useful information without making costly blunders. The software cannot be truly fail-safe but must take individual differences into account.

Subject matter expertise relates to a completely separate body of knowledge, of course. Some highly knowledgeable chemists know little about chemical information searching. The software must more or less lead these users by the hand and at the same time provide many short cuts for the more experienced. However, since the boundaries of science are fluid, a user who is well-versed in chemistry, but far less conversant with biology or physics, may find it necessary to search in all three fields (and others). In the less familiar fields, the user would expect the software to provide aids such as guided searches and thesaurus support. Even another kind of knowledge must be considered, however: a user who knows both the subject area and the methodology of online searching might know nothing about the content and structure of a specific data base. Those who search the CA file with ease, for example, may have more to learn about CASREACT, our reaction data base.

In short, level of expertise is an important aspect of user compatibility, and the standard interface software must be flexible: easy enough for the novice searcher, sophisticated enough for the professional.

Though most scientists worldwide have some facility in English, few non-native speakers have achieved native competence. So even for one who possesses the other kinds of expertise mentioned earlier, language itself can be a critical barrier. The standard software must be compatible with users whose native language is French, or German, or Japanese, or many other languages besides English. For example, if the user knows few synonyms, a good thesaurus is an essential feature. And obviously, all instructions presented in English must be written as clearly and simply as possible.

Subject speciality is in a sense very similar to native tongue, as an aspect of user compatibility. Every subject has a specialized vocabulary, and terminology is used in unique ways in chemistry, biology, physics, and medicine, to name a few important areas. In chemistry, for example, many special characters are used. These are often modifications of familiar Latin or Greek characters and Arabic numbers. But physics and mathematics use many symbols that are foreign to chemistry. Moreover, since the subject matter itself is so different, different kinds of search support programs are required for each discipline.

International Support. As suggested before, the scientific audience is an international one, and the user has a right to expect local support whenever possible. This means getting search questions and software questions answered by a person who is not only knowledgeable about the problem, but also a speaker of the user's language and reachable without a transoceanic telephone call. The user in Germany, for example, should not be required to call the United States to ask about how to link up with the DATEX-P network. Many searching problems may in fact be domestic problems that are best answered locally. Similarly, if the user simply wants replacement copies of lost software or documentation, those items are best delivered by a nearby service center. We believe such support centers are key components of a truly international information service.

Local Edits. This is another necessary feature, and "local" in this case means at the user's computer, as a component of the standard interface software. Those who use word processing software can think of spelling error detection as an example. Just as those programs contain a file of words and spelling rules against which the users can check their spelling, the standard interface software should contain, for example, a file of chemical structure information. This built-up intelligence would be used to determine whether a structure the user has built for online searching is chemically valid. Thus, a chemist could worry less about the details of structure building and devote more time and energy to creative thinking. A variety of other such local editing supports, including a thesaurus and dictionary, would also be invaluable.

Modularity. An additional means of addressing individual differences is to provide the software in modules, i.e., selectable units, because not every user will need the full range of capabilities. One user may have no interest in the chemical structure input capabilities, but a great deal of interest in image display. If these features are separable in the software package, the user may pick and choose. This is a highly desirable option, because it benefits pricing, packaging, adaptability, and sheer convenience from the user's point of view.

Enjoyability. This may seem an inappropriate factor to mention in a discussion of technical services, but we bring it up with no apology. We believe software should be fun to use! Let us consider what that implies.

First, the software should be intuitively pleasing. This means, in other words, it would permit you to find or manipulate information in a way that comes naturally. An interesting paradox of technology is that the more mechanized an operation becomes, the more unnatural it may seem. For example, before word processing came along, you could have "corrected" the transposition of two words by drawing circles around both words, connecting them with a double-headed arrow, and handing the page to your secretary. But performing this simple correction on a computer is not that easy: you must find your place with the cursor, then delete one of the words, then reposition the cursor, and so forth. To become as efficient as the old-fashioned, handwritten process, word processing has to act not only like the pen but also like the secretary.

To most people, it is natural to point at information they wish to select, rather than move a cursor around a screen by pressing a keyboard. It is not by accident that the finger we instinctively use to point came to be called the "index" finger. Thus, it may seem quite logical to the user that he or she could select an item from a menu simply by pointing at it. This fact is closely related to our further observation that software should be tactilely and esthetically pleasing as well.

For example, many of us might prefer to use a light pen or mouse or a similar device as a kind of surrogate index finger in working with a computer. Chemists accustomed to drawing chemical structures on paper may find the mouse more appealing than function keys for inputting structures to their computer. Color is an important component of esthetics, and so even the colors used in the software to highlight or clarify information should not be ignored. Color should clarify rather than confuse, and make searching more enjoyable at the same time.

Since machines are used by human beings, it must be acknowledged that emotive as well as rational considerations determine our tastes in computer interaction to a larger extent than we might believe. These considerations are difficult to separate, however, since the interaction should be ergonomically pleasing too. If the software requires the users to move their eyes all over the screen to perform simple tasks, this not only annoys the user but wastes energy as well. Above all, it must be remembered that using a computer, like most human activities, must be challenging and yet not taxing, to be enjoyable. We believe energy flows from enjoyment and fascination, and we would like to help users produce that energy wherever possible.

Special Difficulties

Quasi-Compatibility. Now that we have considered some desirable features, we must also think about the special difficulties involved in creating a standard interface. In this category, "quasi-compatibility" stands out. If real compatibility is desirable, then false or incomplete compatibility is to be avoided. The problem is that every manufacturer claims the virtue of compatibility whenever possible. IBM clones are plentiful and varied, but we have found many are not truly compatible: just sort of close. We even talked to one user who built his IBM clone himself, only to discover it would not work with truly IBM-compatible software. Ironically, a leading-edge software package, designed to make best use of the IBM advantages, may have more difficulty with the alleged clones than would less ambitious and less useful software. In general, the real professional is more demanding of, and gets more from, machinery than the weekend amateur or thirty-day wonder.

Only a software developer who is well equipped to take full advantage of the capabilities of the operating system can hope to achieve a robust software with true compatibility. One important factor is the compatibility of the software with the way processors and peripheral devices interpret the instructions they receive; different machines obviously interpret instructions in different ways. Thus, for example, the same signal issued by a software program might cause one printer to italicize, another to underline, another to print in boldface, and another to skip to the next page or shut itself off.

Incompatibilities can surface in display adaptor and device driver differences and affect the performance of Video Displays and printers. To create software that takes the differences into account, programmers must have access to all the equipment involved. The hardware and software must be tested thoroughly at many points along the way: the software developer must have the resources to provide this array of test equipment and commitment to thoroughness. Consider modems, for example. Since the Hayes modem is the standard, virtually any manufacturer would claim to be Hayes-compatible. But if the software developer takes the manufacturer's word at face value, he might discover from a customer that the modem does not actually work with his truly Hayes-compatible software.

Different Worlds. Each computer manufacturer presents a "different world" to the user, and the different worlds presented by IBM and Macintosh deserve special consideration. Consider the mouse, for example. In the Macintosh world, the mouse has one button; in the IBM world, at least two. In the Macintosh world, the icon orientation has been evident from the beginnning, with functions represented by pictures such as trash cans and clipboards; in the IBM world, menus still dominate, for the most part. A responsible software developer will write programs that take advantage of the best features of both worlds. This is very expensive and a special difficulty encountered in the creation of a standard interface. For one thing, the developer must avoid losing the advantages of one machine by trying to exploit the advantages of two. This would be analogous to putting a Yugo engine into a Lamborghini; you could probably make the hybrid vehicle work, but would anyone want to drive it?

Special Protocols. Special protocols occur on each VAN and online vendor system, and on the corporate information systems as well. The special protocols require different log-in procedures, prompts, and timings. Handling all the different error conditions that might occur increases the software developer's difficulty by an order of magnitude, at least. Conditions such as "invalid user ID", "invalid address", and "system busy" apply to all the networks and handling them may require many lines of code.

Transmission. Other difficulties involving networks include transmission speed and noise. The software currently must work at 300 baud, 1200 baud, and 2400 baud to serve the variety of user equipment available. In the very near future, 9600 baud transmission will be commonplace, and new and better means of communication will be the norm. Noise and transmission errors must be dealt with, to be sure that the data are delivered both ways in good condition. We have found that even the error-correcting protocol Kermit is not 100% reliable, 100% of the time.

Cost Versus Price. Finally, cost versus price is a special difficulty in the software field. The complexity of the front-end software is a major factor in the cost, in that more complex software needs to be more highly priced. It takes more work and expertise to develop and this makes it more useful. But, ironically, this complexity may not be apparent to the user especially in the best designed and most serviceable software! By analogy, consider that when a system of the human body works well, you tend not to notice it. STN Express software is actually quite complicated; programs that appear simple in their execution may have cost the developer months of programming, many thousands of lines of code, and hours of design and prototyping. And this must have some effect on the price.

Another major influence on price is the size of the market for scientific software. For software that serves a mass audience, the price can be relatively low. But STN Express is expensive to produce and of interest to only a limited audience, probably measured in the tens of thousands. So the price must necessarily be relatively high.

Conclusion

Standards are necessary to facilitate communication in scientific information retrieval, as in all other areas of human activity. Therefore, a standard interface software package that accommodates users searching public, corporate, and personal files is highly desirable. Setting the standard is a perilous task because of differences that exist among networks, operations and applications software, hardware, the kinds of data that must be retrieved, and the needs and preferences of users. With the future in mind, we must take care not to set standards too limited to adopt to new scientific discoveries and even new modes of thought. Compatibility is the *sine qua non*, and the first and foremost feature of a standard interface software package is to be compatible with the user. This requires the software developer to pay careful attention not only to harmonizing the various mechanical components of searching, but also to planning for the varying levels of expertise and special interests of the users.

Bridges must be built, if the standard interface software is to communicate with many online vendors and permit the user to integrate retrieved information with specialized software, such as word processors, statistical packages, and many other types. We believe the association of CAS and Evans and Sutherland is a good example that will open new opportunities for integrated chemical structure searching and molecular modeling. Other key alliances between hardware manufacturers and software developers alike remain to be forged. Different machines and programs present the user with different worlds, and more cooperation can make the best of all worlds possible.

RECEIVED May 2, 1989

Chapter 7

Towards the Universal Chemical Structure Interface

William G. Town
Hampden Data Services, 167 Oxford Road, Cowley,
Oxford OX4 2ES, England

Chemists are trained to communicate chemical information using graphical images, namely, chemical structure diagrams. In consequence, they have been among the earliest and most demanding users of graphical interfaces to computer systems. In order to be able to pursue their research and develop new ideas, they need to conduct searches which combine structural concepts with text and/or data concepts. As most chemists are prepared to learn to use only one or two chemical information systems, the same graphical chemical structure interface should give access to personal, company and public data bases and should also be an integral part of chemical word processing software, chemical CD-ROM products and laboratory microcomputer data bases. The PSIDOM chemical structure drawing interface has already been integrated in a number of products including STN Express, a PC-based front-end package; TORC, a Derwent Ring Code fragment generation program; and Sadtler's IR spectra library search software. All are compatible with the integrated PSIDOM range of software.

In today's research laboratory, the largest growth in computing power is occurring not in the central computer facility but on the research worker's desk in the form of personal computers and, increasingly, personal workstations. The availability of local computer power, local data storage and a large bandwidth between the user interface and the processor is dramatically changing users' perception of computers and their application software. Derivations of software environments originating in the Xerox Palo Alto Research Center (PARC) have evolved at Apple Inc into the Apple Macintosh Human Interface, and have also been implemented in other hardware and software environments (e.g., IBM's OS/2 Presentation Manager; Microsoft Windows; GEM, Graphics Environment Manager, from Digital Research; Massachusetts Institute of Technology's X-windows, etc.). These *de facto* standard environments are rapidly changing users' expectations of software design. Today's user interface is increasingly graphically based, using windows which drop down, pull down or pop up, icons for the representation of files and modes of operation, and a mouse or other pointing device to move a cursor around the screen and to select objects or commands. This increasing standardization of user interfaces is reducing the difficulty experienced by users in moving from one application

0097–6156/89/0400–0068$06.00/0

software package to another and thereby increasing users' productivity as more tools become accessible.

However, a consistent user interface is only one of several equally important aspects of today's research computing environment. That environment is increasingly a distributed computing system in which connectivity, integration and open architecture are equally important to the software designer and the user alike. The term "connectivity" addresses the communications and other issues relating to the interfacing of and transfer of data between applications running on personal computers and mainframes. "Integration" implies a higher level of connectivity in which the relative personal computer and mainframe applications have been designed to work closely together. "Open architecture" is an approach to software design which allows and encourages integration of software components from different producers by the use of well-documented, modular interfaces to system components.

When the personal computer revolution began in the early 1980s, personal computers were small by all usual measures; they had small random access memories (RAM), their permanent storage capacities were small (typically a 360 Kbyte floppy disk) and their processors were slow. Today, two generations of personal computer design later, a personal computer may have several megabytes of RAM, several hundred megabytes of permanent storage on hard disk and processors capable of several millions of instructions per second (MIPS) making them comparable in power to mainframe computers of only a few years ago. New operating systems offering multitasking and multiprocessing are also enhancing their capabilities dramatically. Applications software packages (such as molecular modeling and computer aided design), which are highly dependent on graphics display and which are computationally demanding, have already migrated to personal workstations whose cost, with the ever falling cost of hardware components, has reduced to a level comparable to that of a top of the range personal computer. This blurring of two previously distinct concepts leads to the interchangeability of the terms "personal computer" and "workstation" and both are used freely in this chapter to refer to basically the same concept.

Other types of applications which are dependent on access to a centralized data base (e.g., information retrieval) or requiring the sharing of information between users (e.g., office automation) implicitly require solution of problems such as connectivity and open systems architectures before they can be successfully implemented as applications in distributed systems. In such a system the user should be able to access and process data through a consistent user interface whether the data is stored in a local file, a centralized company file or a publicly available file and be able to move data and queries seamlessly from one application package to another. The ease of transporting data and queries between applications increases if standard exchange formats and standard data representations are available but open architecture becomes a reality only when software producers are compelled to adopt these standards as a result of pressure from their users.

Chemists' Computing Requirements

The chemical structure diagram is an important component of the natural language of chemistry. Any two chemists talking about chemistry will quickly reach for pencil and paper or blackboard and chalk and start sketching structure diagrams to communicate and clarify the chemical concepts they are discussing. Most scientific articles and reports in fields relating to chemistry are peppered with chemical structure diagrams. It is hardly surprising therefore, that many of the application software packages used by chemists include the input and display of chemical structure diagrams. In consequence, chemists have been among the earliest and most demanding users of graphical interfaces to computer systems. Among the applications software packages used by chemists, diagrams form an important communication mechanism in chemical structure retrieval (i.e., structure and substructure searching), reaction retrieval, synthesis planning, molecular modeling, structure elucidation, quantum chemistry calculations and in report generation. Add to this impressive, but incomplete, list the requirement to communicate structure diagrams through messaging systems and provide seamless access to local, company and public data bases and it is immediately apparent that the design requirements for the "Chemist's Workstation" are demanding.

A schematic representation of how a chemist's workstation may fit into a distributed research computing system is shown in Figure 1. The diagram is, of course, a dramatic over-simplification of the real world. Several comments are needed before moving on to consider some of the complexities which might prevail. Firstly, the chemist's workstation becomes his window on the whole of his computing environment. Initially, it must provide for emulation of "dumb" terminals (both alphanumeric and graphic) for, until the software running on mainframes within companies and public hosts universally adapts to distributed

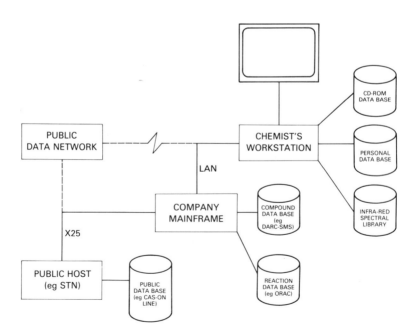

Figure 1 The Chemist's Workstation

system environments, this now outmoded method of interaction with application software must be supported. Secondly, it could provide at the local level many of the functions which are also supported at the central level, including word processing, data base management, text and structure retrieval. Thirdly, it must provide seamless access to the data bases supporting these functions whether stored locally or centrally.

Local Data Bases

Figure 1 shows personal data bases stored at the chemist's workstation which may have been created by the chemist from his own input or from information downloaded from company or public data bases, which, when placed in a certain juxtaposition or annotated with the chemist's own comments or data, may take on new meanings. The diagram also shows a CD-ROM data base to represent one of the many forms in which commercially available data bases may be provided to the chemist for his use in the near future. Such data bases will replace, in part, the small library of handbooks which have a place in the chemist's laboratory or office. An example of the type of data base becoming available for use by the chemist is the Standard Drug File published by Derwent Publications Ltd containing information (including chemical structure diagrams) on approximately 17,000 chemical compounds found in the pharmaceutical literature. This data base, created by Hampden Data Services (HDS) for Derwent, will first be made available as a structure searchable file for personal workstation use with HDS's PsiBase software (a component of the PSIDOM range).

Laboratory Data Bases

The infra-red spectra library is included in Figure 1 to represent laboratory data files used in chemical analysis or structure elucidation which again may have been created locally or purchased either individually or as part of a laboratory instrument computing system. Sadtler Research Laboratories, a Division of Bio-Rad, have recently released a new version of their infra-red spectra data base containing information on some 80,000 compounds including searchable spectra and structure information. Again the files are available for use on the chemist's workstation and are searchable by PsiBase which is marketed as the Sadtler Substructure Search Software. When fully developed the system will allow searches by spectral characteristics and substructures in an integrated manner within a Microsoft Windows environment. Other types of spectral data bases to be added to the system include nuclear magnetic resonance (NMR) and mass spectra files. VG Masslab, a company which manufactures and markets mass spectrometers, are building, with the assistance of HDS, a structure-searchable version of the widely used National Institutes of Health/National Bureau of Standards/Mass Spectra Data Center mass spectra library which now contains some 50,000 compounds. Finnigan MAT, another mass spectrometer manufacturer, is developing a new version of its software package known as the ChemMaster workstation which will integrate spectra from a variety of sources and make them structure searchable with HDS's PsiBase software.

Company Chemical Structure and Reaction Data Bases

During the last ten years, chemical and pharmaceutical companies' research divisions have compiled, and made accessible to their research workers, large data bases of chemical structure and chemical reaction information which are stored in centralized systems. The software systems which provide access to these centralized data bases (MACCS, DARC-SMS and OSAC for structures; REACCS and ORAC for chemical reactions) are, in their present versions, not fully adapted to distributed software environments. The CPSS software from Molecular Design Limited (MDL) does allow the user to access and use data from MDL's MACCS and REACCS products within ChemText and ChemBase (which are modules of CPSS). Structures and reactions created with HDS's PSIDOM software can be uploaded into ORAC Ltd's OSAC and ORAC software and into Télésystèmes' DARC-SMS software. Experiments with one client have even proved the feasibility of transferring PsiBase queries to a MACCS environment. Nevertheless, at present, it cannot be said that the desired open architecture integration and transparency between personal and company data bases exists. However, collaboration between software providers, data base producers and major information users in the development of a standard exchange format for chemical structure related information gives hope for the future. The development of the Standard Molecular Data (SMD) format is described in Chapter 11 of this book.

Public Data Bases

The recent development of PC-based front-end packages such as STN Express, MOLKICK, DARC-CHEMLINK and ChemConnect has eased the use by chemists of public online data bases, such as CAS ONLINE and Beilstein. All four packages enable the chemist to prepare queries for chemical structure searching whilst working offline on his personal computer. STN Express includes query validation routines which alert the user to possible problems which may occur in a search before the search is executed and has the CAS rules for normalization (i.e., aromaticity and tautomerism) built into the software. One other major feature of STN Express is the communications package built into it which allows the transmission of the query connection table to the host to be performed under an error checking protocol, thus ensuring that the query arrives at the host computer in good shape. Two of the other three packages rely on the use of standard communications software. Finally, a potential advantage for the chemist is the use of the same chemical structure drawing interface in PsiBase and STN Express and the upward compatibility of queries between PsiBase and STN Express enabling the same query to be searched against a local structure data base and either CAS ONLINE or Beilstein on STN International's host computers in Columbus or Karlsruhe.

The PSIDOM chemical structure drawing interface is being used in two other products which facilitate searching of online chemical information. Both of these products have been commissioned from HDS by Derwent Publications Ltd and are aimed at searching text-based files which contain chemical fragment codes used as index keywords. Until recently, the user of these files, covering the pharmaceutical literature (RINGDOC) and chemical patents (WPI/L), had manually to encode the chemical structure to be searched using the appropriate fragment code (the codes used are quite different in the two files). Now it is possible to draw the chemical structure and have the codes generated automatically by the appropriate software package: TORC (TOpological Ring Code generator), for the RINGDOC file, and TOPFRAG (TOPological FRAGment code generator) for the WPI/L file. TORC incorporated the PSIDOM chemical structure drawing interface from the beginning and a new version of TOPFRAG with the same interface is under development. This new version will cope with generic (partially undefined) queries as well as specific chemical structures.

Inclusion of Chemical Diagrams in Documents

Scientific documents are complex entities. They are built up from a number of elements including text, graphics, tables, chemical structures and equations. Each presents its own particular problems when considering the stages a document may enter during its life cycle (i.e., preparation, publication, dissemination, archival storage and subsequent retrieval) and each stage may be further complicated by differentiation into its paper and electronic variants. The topic is too complex to cover in any depth here so rather I will concentrate on the integration of chemical structures with text.

Graphics for chemical structures have already been covered by an ACS Symposium Series monograph (*1*) and one of the most active areas of software development in recent years has been for the integration of chemical structures with text and data (*2*). In view of the excellent reviews of the software available in this area, it is not necessary to catalog the available packages and their relative merits or demerits. Rather, I will try to present the general approaches which are possible and the advantages or disadvantages of each.

Chemical structures may be represented, *inter alia*, as character matrices, bit-map images, line art or connection tables, the latter representation being the only one in this list which is suitable for structure searching. Fortunately, connection tables are suitable for generating chemical structure diagrams for display and printing and they may easily be converted into line art or bit-map images. They represent a highly compact form of representation, a typical connection table occupying a few hundred bytes compared with several thousand bytes for bit-map images, a fact which is of great significance when data bases of several thousand chemical structures are to be created. Connection tables have one additional advantage over other representations in that they are easily modified with a suitable editor (such as the PSIDOM chemical structure drawing interface as implemented in PsiGen) which may incorporate chemical intelligence.

The integration of chemical structures with text may be achieved in various ways according to the type of representation in use. The character matrix approach was one of the earliest solutions to creating structure diagrams within scientific word processors by the use of special characters to represent line and angle segments. The results are often unsatisfactory for bridge ring and other three-dimensional structures. The second approach is to achieve text and structure merging by importing text or print files from word processors into the "integrator" package which may incorporate a structure editor but not usually a text editor. Most chemists have a preferred word processor and the trend is for "standard" word processors (such as WordPerfect, and Lotus Manuscript) to accept line art in the form of graphic metafiles (usually the ANSI standard Computer Graphics Metafile, CGM) which may be exported from a structure editor such as PsiGen. Once imported into the word processor the diagrams may be scaled but the chemistry may not be changed. Some chemists prefer to use a product such as Lotus Freelance to enhance diagrams imported from a structure editor before creating presentation graphics or before including them in reports. For reports and scientific papers where publication quality is required, the same metafiles may be used in desktop publishing packages (such as Xerox's Ventura Publisher) which also accept text files from many of the standard word processor packages.

Electronic Transmission of Chemical Reports

Increasingly, transfer of information between researchers in the same company is occurring *via* electronic messaging or mail systems – often as a component of an electronic office environment (e.g. DEC's All-in-1 or IBM's Professional Office System PROFS). As discussed above in the context of document preparation, the need also arises for the inclusion of chemical structure diagrams in reports and other documents which are to be transmitted electronically. It is not sufficient generally to provide a passive, read only, version of a document since the recipient may need to edit the text and the diagrams included in the report. This is usually not a problem for the textual component of a document as the electronic office environment is usually homogeneous and contains a built-in word processor. If the structure editor is standardized throughout the company the optimal solution is to "attach" the binary files created by the structure editor to the document and transmit the whole to the recipient. If this is not the case, or if the mail system does not cope with binary files, the representation must first be converted to an ASCII representation, such as the SMD format, before transmission.

Conclusion

The ideal of a universal chemical structure drawing interface has not yet been achieved but, through the adoption of the PSIDOM interface by a variety of software and data base producers, significant progress has been made in this direction. As new versions of existing applications software packages emerge for the Macintosh and OS/2 Presentation Manager platforms, the productivity of the user will increase following the simplification these *de facto* standard interfaces will bring to the task of using new applications. Collaboration by software and data base producers on the SMD format will simplify data exchange and may lead to open architecture systems in which the chemists workstation is truly his window on the whole of his computing environment.

Literature Cited

1. *Graphics for Chemical Structures: Integration with Text and Data*; Warr, W.A., Ed.; ACS Symposium Series 341, American Chemical Society: Washington, DC, 1987.
2. Love, R.A. In *Chemical Structure Software for Personal Computers*; Meyer, D.E.; Warr, W.A.; Love, R.A., Eds.; ACS Professional Reference Book, American Chemical Society: Washington, DC, 1988.

RECEIVED May 2, 1989

Chapter 8

A Universal Structure/Substructure Representation for PC-Host Communication

John M. Barnard
Barnard Chemical Information Ltd, 46 Uppergate Road, Stannington,
Sheffield S6 6BX, England

Clemens J. Jochum and Stephen M. Welford
Beilstein Institute, Varrentrappstrasse 40–42, D-6000 Frankfurt 90, FRG

The development of PC-mainframe communication programs for chemical structure searching is discussed. The configuration of the MOLKICK and S4 programs is outlined, and the ROSDAL string format, used to transmit chemical structure representations between PC and mainframe, is described. ROSDAL is an ASCII character string, which can be used both for uploading of queries and downloading of retrieved file structures. Its simple syntax allows both ease of automatic interconversion with other representations, and manual encoding and decoding.

A number of systems now offer graphics-based substructure search capabilities for online access to a variety of chemical structure data bases. From December 1988 the first subset of the Beilstein Online data bank (organic heterocyclic compounds reported in the literature prior to 1960) goes online on STN International. Other hosts, which at present offer only name-based structure and substructure search capabilities are also likely to provide online access to the Beilstein files in the near future; these include Dialog, Pergamon Orbit Infoline and Datastar.

PC-Based Terminal Emulation Programs

A variety of programs can be used to enable a PC to emulate an ASCII terminal. These are familiar to all PC users who use their PCs for online searching. Certain of them, listed in Figure 1, provide only non-graphics emulation, while others (Figure 2) also provide graphics emulation (1). These latter support Tektronix or similar graphics standards and protocols and enable the PC to input and transmit and receive and display graphics images, including chemical structure diagrams, to and from an online host. The emulation programs may themselves contain PC-host communication software (e.g. PC-Plot) or if greater capabilities are required, such as automatic log-on, downloading, file management, etc., they can be used in conjunction with specialist communication programs (e.g. Crosstalk).

0097–6156/89/0400–0076$06.00/0
© 1989 American Chemical Society

Crosstalk
Vterm
Mirror II
Dialog-Link
(Symphony)
(Framework)
(MS-Windows)

Figure 1. Terminal Emulation Programs. No graphics emulation

Emutek
PC-Plot
TGraph

Figure 2. Terminal Emulation Programs. Graphics emulation (Tektronix 4010/4014)

Front-End Programs for Chemical Structure Searching

Even if such emulation programs are used, however, graphics interaction with a host has, until recently, required mainframe graphics software to create and transmit the graphics images. This has been the case with, for example, CAS ONLINE (*2*), DARC (*3*) and, for in-house systems, MACCS. This requirement has had the disadvantages of being slow and prone to transmission errors since the quantity of information transmitted is high, as well as expensive since the transmission time is correspondingly great.

Offline preparation and presentation of chemical structure graphics is now possible on a PC using a variety of query editor programs (*1*), such as those listed in Figure 3. In most cases these are specific to a single system, and use a proprietary format for transmission of the query structure between the PC and the mainframe computer. For example, STN Express allows query structures to be built offline and then, after connection to STN, uploaded and searched against the STN online structure files. Retrieved structures can be downloaded onto the PC and browsed offline in STN Express. Similarly, CHEMLINK is now available for Télésystèmes-DARC, while in the case of in-house systems, ChemBase provides similar capabilities for MACCS.

STN-Express
DARC-Chemlink
BEILSTEIN-MOLKICK

Figure 3. Chemical Query Editor Programs

MOLKICK and S4

The MOLKICK program, now available from Softron GmbH is collaboration with the Beilstein Institute, provides a wide variety of query-drawing functions, templates and shortcuts. Although it has been developed primarily for use with the newly-developed Softron Structure/substructure Search System (S4), unlike the other, system-specific, query editor programs, MOLKICK is able to create a suitable query definition for uploading to other search systems (4), such as STN or Télésystèmes-DARC.

The S4 search system itself embodies several novel and interesting concepts. The search file is built from the Beilstein structure file by encoding every atom of every molecule in terms of its extended connectivity encompassing every atom in the molecule. These compact atom codes are sorted into a file and further compressed. From this large file of atom codes a search tree is generated consisting of the first ten spheres around each atom. This tree forms the index through which the atom code file is accessed.

The most discriminating code which encompasses all relevant aspects of the query structure is generated from the query connection table. A search for this query code in the index results in a list of addresses in the atom code file. The atom code file is then accessed and read from the starting address until the coding changes. Contained in this list are the hits; in most cases no atom-by-atom search is necessary and, because of the minimized number of disk accesses, very fast search times can be achieved. For this reason, S4 is ideally suited to searching very large structure files and it is presently being extended for stereochemical search and for tautomer search.

Transmission of Chemical Structure Queries

To minimize network transmission and online connect costs, chemical structure and substructure queries and the resulting answer structures should ideally be sent in a heavily compressed format. On the other hand, chemists and information specialists not equipped with PCs should be able to generate this format manually and verify answers.

When communicating with the S4 program, MOLKICK transmits and receives chemical structures in the form of a string of ASCII characters, called a ROSDAL string (Representation of Structure Diagram Arranged Linearly). Figure 4 illustrates the overall system by which MOLKICK communicates with S4, using ROSDAL strings. The query structure or substructure is converted into ROSDAL format and transmitted to the host computer. On the host computer, S4 converts the ROSDAL query string into its own connection table format, and executes the search. Retrieved structures are converted into ROSDAL format and transmitted to the PC where MOLKICK reconverts the ROSDAL string into a graphics image for display.

ROSDAL String Format

Figures 5, 6 and 7 illustrate the ROSDAL strings for some example structures. In the string each non-hydrogen atom is arbitrarily but uniquely numbered, and the string consists of several sequences of connected atoms, separated by commas. Each atom is identified by its number, which may optionally be followed by its element type (if different from carbon) and by other symbols in parentheses, giving information on its charge, mass, stereodescriptors etc.; additional node attribute symbols can be used to indicate free sites ("*") and attachment points ("&").

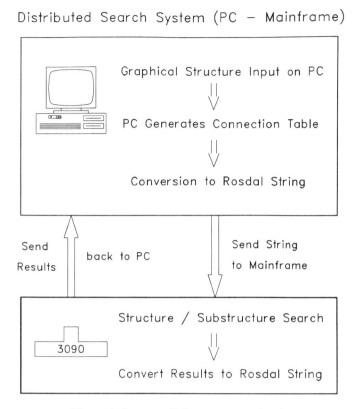

Figure 4. Structure/Substructure retrieval

Bonds are shown by symbols indicating single, double and triple bonds, which are placed between the atoms, with optional additional information in parentheses giving bond orientation (e.g., Above or Below the plane), stereodescriptors, or ring or chain environment (for query substructures).

There may be any number of sequences of connected atoms in a ROSDAL string; at one extreme every pair of connected atoms can be shown in a separate sequence, whilst for many structures a single sequence can encompass virtually all the atoms and bonds. It is not necessary to show hydrogen atoms explicitly, though they may be shown if desired: in this case they must also have unique numbers.

A "shortcut" notation can be used for chains of consecutively numbered atoms, for example in a ring system, in which the starting atom is followed by two bond symbols, and then the finishing atom; this is illustrated in Figure 7.

The syntax of ROSDAL is formally defined by a set of Backus-Naur Form production rules, which are given as Appendix C to the MOLKICK User Manual (5). The simple syntax facilitates both automatic processing and, if desired, manual encoding and decoding. Because it is entirely composed of ASCII characters, the string may be edited using a text editor. ROSDAL strings are unambiguous but nonunique descriptions of chemical structures, and many equivalent ROSDAL strings can be built for a single structure.

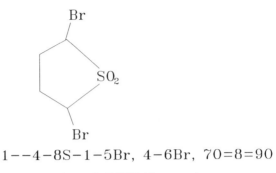

1--4-8S-1-5Br, 4-6Br, 7O=8=9O

Figure 5. ROSDAL example

Molecule: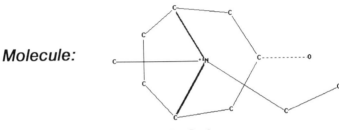

Rosdal String:

1N(+1)-5,1-4-10,1-(A)3-9-7-2-6-11-(B)12O,1-(A)2,3-8-11.

Figure 6. ROSDAL example

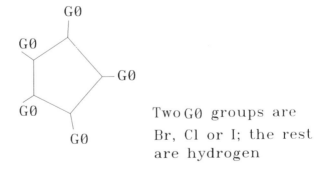

Two G0 groups are Br, Cl or I; the rest are hydrogen

1--5-10G0, 2-6G0, 3-8G0, 4-9G0,
5-7G0;G0=2* (1Br&; 2Cl&; 3I&)

Figure 7. ROSDAL example

Though it was developed completely independently, ROSDAL has certain features in common with the SMILES notation (6), though unlike SMILES, ROSDAL is not involved in the search process. ROSDAL is able to represent both full structures (including stereobonds), and substructure queries incorporating features such as free sites. In addition, it has the capability of representing a number of generic query structure features, including special generic atom types, generic group nodes, and alternative substituents (as shown in the example in Figure 7). Its formally defined syntax, which is analogous to that of the GENSAL generic structure description language (7), developed at Sheffield University, may also allow it to be used in conjunction with GENSAL for the transmission of more complex generic structure representations.

The ROSDAL syntax is also currently being extended to accommodate inorganic substances. Details of these developments and other aspects relating to structure representation in the Beilstein structure file can be obtained from the authors.

Acknowledgments

This work was partially supported by the German Ministry for Research and Technology. The authors would like to thank Professor M.F. Lynch (Sheffield University) and Dr Peter Jochum (SOFTRON GmbH) for helpful discussions.

Literature Cited

1. *Chemical Structure Software for Personal Computers*; Meyer, D.E.; Warr, W.A.; Love, R.A., Eds.; ACS Professional Reference Book; American Chemical Society: Washington, 1988.
2. Farmer, N.A.; O'Hara, M.P. *Database* **1980**, *3* (4), 10–25.
3. Attias, R. *J. Chem. Inf. Comput. Sci.* **1983**, *23*, 102–108.
4. Hicks, M.G. In *Software Entwicklung in der Chemie 3*, Proceedings of the Workshop *Computer in der Chemie*, Gauglitz, G., Ed.; Springer-Verlag, *in press.*
5. MOLKICK Users Manual, Copyright 1988, SOFTRON GmbH.
6. Weininger, D. *J. Chem. Inf. Comput. Sci.* **1988,** *28*, 31–36.
7. Barnard, J.M.; Lynch, M.F.; Welford, S.M. *J. Chem. Inf. Comput. Sci.* **1981**, *21*, 151–161.

RECEIVED May 2, 1989

Chapter 9

Chemical Structure Information at the Bench: A New Integrated Approach

Robert M. Olszewski, Everett A. Bruce, Craig Leilous, Rudy Potenzone, Jr.
Polygen Corporation, 200 Fifth Avenue, Waltham, MA 02254

Traditional systems currently available do an excellent job of providing very specific solutions to specialized aspects of a researcher's information needs. These include structure searching, data retrieval, data graphing, analysis, modeling, etc. However, in observing a working chemist, it is clear that this is only part of the problem. The generation of documents which integrate aspects of all these applications is required. These documents include (but are not limited to) technical reports, research progress reports, scientific papers and laboratory notebooks. Easy generation of these documents is essential to the day to day effort of the chemist. However, from the research organization's perspective, future access to these documents, as well as the underlying data, is also crucial. The CENTRUM system is an integration tool to address these needs. The general system architecture is described along with some of the advanced features.

Chemical and pharmaceutical companies are continuously searching for ways to decrease the time taken to bring new products to market while reducing the costs associated with development. As part of this effort, scientific computing groups and senior management are reexamining how computers may be used by researchers to increase effectiveness and efficiency during the discovery process. Two areas of concern are access to chemical information and the final presentation of research data in documents. Certainly the creation and accessibility of chemical information have been improved through the use of computer data base and chemical structure drawing software. But these applications were developed as stand-alone solutions, and can only be loosely integrated. Interchange and merging of these scattered information sources can often be awkard or impossible to achieve. All too often chemical and pharmaceutical organizations are forced to pattern their research methodologies according to the available computing tools, thus diminishing the benefits of a computerized environment. This paper discusses CENTRUM a modern solution for improving access and extraction of chemical information, data analysis and finally the presentation of the information in the form of journal articles, technical reports or professional publications.

Chemical and pharmaceutical research organizations generate, analyze and distill large volumes of scientific data. Laboratory notebooks track daily research progress, accumulating data in the form of structures, equations, reactions, tables, graphs and text. Tables are commonly used to contain information relating to compounds and selected characteristics. All of this information is organized and eventually published.

Recent developments and innovations in computer software and hardware are altering the ways research professionals and management use the corporate information base. The emergence of graphics and user interface standards coupled with the introduction of low cost, high performance 32-bit workstations now make it possible to build a highly integrated research environment. The Massachusetts Institute of Technology's (MIT) X-Windows (see Figure 1) System (funded by thirteen hardware and software companies including IBM and Digital Equipment Corporation) defines an industry standard for graphics and network computing. This system gives software developers a universal application interface, independent of hardware or operating system. With consistent engineering it is possible to port applications to any system supporting X-Windows. The X-Windows system also establishes an industry standard for distributing complex computing tasks across a network. It lets applications execute on a networked host (client) and display on a desktop device (server) such as a workstation (VAX, Hewlett-Packard, Sun, or IBM), an IBM PC AT or a Macintosh running X-Windows server software, or an X-Windows terminal. The Open Software Foundation's (OSF) industry standard user interface, called MOTIF, provides applications with a consistent user interface on any X-Windows hardware platform. Finally, hardware vendors have introduced high-powered 32-bit desktop workstations at prices which are competitive with high end PCs. As a direct result of these developments, the scientist and researcher will have a set of highly integrated applications programs with the same "look and feel" available on a wide variety of computers. The research organization will be able to connect computers that span a wide price and performance ratio, allowing companies to take advantage of their investment in existing PC ATs and software applications.

Overview of CENTRUM

In anticipation of these developments Polygen has designed a product which makes best use of industry standards and "state of the art" computing platforms. The CENTRUM research workbench is composed of two domains, chemical publishing software, and information handling software, specifically designed to meet the needs of scientists, technical support staff and research management. The CENTRUM Chemical Publishing Environment (CPE) contains a number of chemical and scientific object editors creating chemical and scientific graphics (structures, reaction diagrams, graphs, equations and compound elements such as tables). The CPE graphics are then inserted into the text using the graphics integration tools provided by the word processor or publishing package. This process can be used to enhance internal documentation, technical reports and journal articles. The CENTRUM Chemical Information Manager (CIM) provides the tools necessary to access, retrieve and organize information from data processing applications running on networked hosts. Both CPE and CIM applications run on a wide variety of computers, providing a flexible software environment designed for growth and integration.

Polygen's CENTRUM Release 3.0 embodies an "open systems" software platform that allows research organizations to integrate existing third party

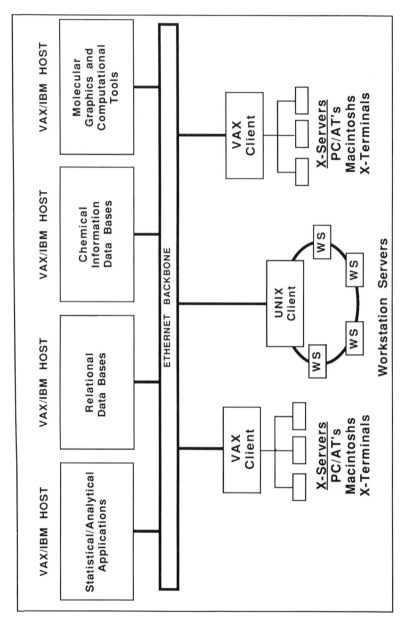

Figure 1. Distributed Application Computing Environment

software or build custom applications to meet their computing needs. CENTRUM software serves as the basis for the researcher's workstation, without bias toward a particular hardware platform. CENTRUM's modular design allows information access and publishing solutions to be installed independently or together to create a tightly coupled arena for information access, retrieval and document preparation.

Chemical Publishing Environment

CENTRUM offers unparalleled integration. The chemical objects (structures, reaction diagrams, graphs, equations and tables) can be integrated with any document processing or publishing package that accepts industry standard picture formats such as Postscript (the page description language from Adobe Systems), PICT (Apple's storage format for graphics images), CGM (the ANSI standard Computer Graphics Metafile), DDIF (Digital Equipment's Digital Document Interchange Format) and MIF (Frame Technologies' Maker Interchange Format). CPE editors share a common user interface and produce high quality graphics to represent the associated scientific data (see Figure 2). For example, CPE's chemical structure editor is used to sketch quickly 2-dimensional chemical structures. Every structure created with the editor not only has a picture file but a connection table which describes how the various atoms and bonds are combined. This means that the creation and editing of structures can be accomplished in a way that makes chemical "sense", e.g., valences are satisfied. Furthermore it means that the results of the chemical sketching program can be used by other chemical applications, such as 3-dimensional molecular construction, structural analysis or input to chemical modeling applications. CPE complements existing word processing software, eliminating the disruption and retraining of staff and the costs associated with archive document conversion.

The ability to integrate chemical graphics with a wide range of text processors is an attractive advantage of Polygen's Chemical Publishing Environment. In some cases end-users may opt to use advanced publishing packages which provide a higher level of integration and conform to the industry standard definition for compound documents, for example, Digital Equipment's Compound Document Architecture (CDA), a widely endorsed standard which defines a methodology for integrating text, graphics and data. FrameMaker from Frame Technologies Inc. and DECwrite from Digital Equipment are two examples of publishing packages which integrate CENTRUM graphics and data. These packages support "hot links" to the chemical information and data underlying the CENTRUM picture file. The hot links provide access to CENTRUM applications directly from the document. When CENTRUM CPE is used with either FrameMaker or DECwrite the user can access the CPE editors without leaving the document composer, simply by pointing at a CENTRUM graphic and clicking a mouse button. The "hot link" mechanism automatically invokes the appropriate CPE module. Upon exiting the CPE application the updated graphic is returned to the document.

Chemical Information Manager

Polygen's Chemical Information Manager (CIM) provides research organizations with a new class of information management tools. Unlike other information

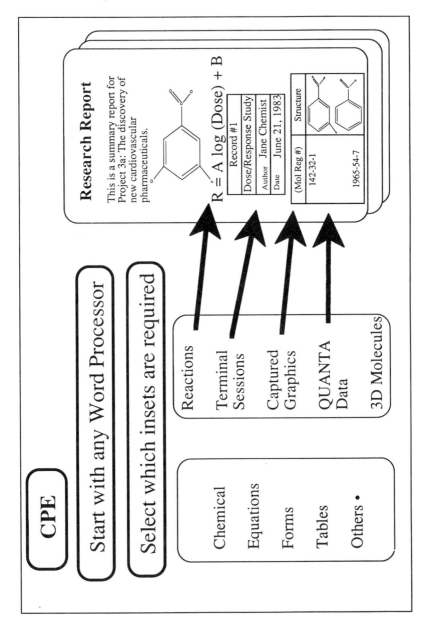

Figure 2. Chemical Publishing Environment (CPE)

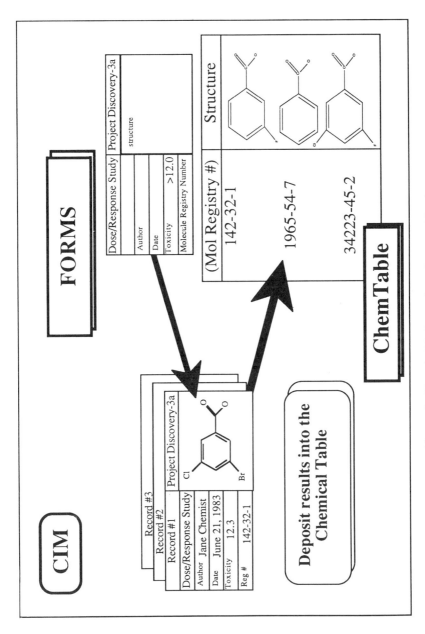

Figure 3. Chemical Information Manager (CIM)

management applications, CIM accesses, retrieves and merges chemical information from multiple structural and relational data bases with a single user interface. CIM enriches the content of gathered information by forging additional connections to related information. Research organizations can also build their own applications that contain tailored data access rights for all levels of the organization. Training and support requirements are simplified since the Chemical Information Manager frees users from the burden of learning complex network and application start-up commands.

CENTRUM CIM offers several information handling tools (see Figure 3). FORMS provides the functionality required for forms-based data entry and retrieval. ChemTable possesses formatting and data analysis capabilities and integrates output from all CENTRUM CPE modules. ChemTable lets research groups build, maintain and manage complex information relationships. Employing a spreadsheet-like format, ChemTable organizes collected information into rows and columns. Statistical and analytical functions may be performed on information contained in the cells. ChemTable allows the user to take advantage of the processing power of external third party applications without exiting the chemical spreadsheet environment.

Using Polygen's Scripting Language (PSL), an advanced macro scripting language, research groups can generate their own ChemTable applications. PSL can be used to automate repetitive procedures performed on ChemTable data or other CENTRUM objects. PSL can also be used to link data contained in ChemTable to external applications.

Summary

The CENTRUM research workbench represents the next generation of chemical information processing tools. CENTRUM CPE and CIM are the first applications that take advantage of the established industry standards for document, graphics and network communication. CENTRUM's conformance to industry standards guarantees compatibility with software solutions from other vendors observing the same standards. CENTRUM works with standard network protocols and data definitions and provides a consistent user interface. These design criteria result in CENTRUM applications that build upon the basic connectivity offered by systems hardware and software vendors, providing unparalleled uniformity of service for chemical publishing and information access.

RECEIVED May 12, 1989

Chapter 10

Building a Comprehensive Chemical Information System

Jean-Pierre Gay, Guillaume Auneveux, and Françoise Chabernaud
Télésystèmes, 83–85 Bld Vincent Auriol, 75013 Paris, France

Building a comprehensive chemical information system today requires the integration of several different types of software. For complete integration, the chemical structure software must be linked with various mainframe data base management systems (DBMS) such as ORACLE, 1032, DB2, or DS/SQL, and should also link with various personal computer (IBM PC or Macintosh) software packages. This paper will describe three software packages which have been developed by Télésystèmes in cooperation with several European pharmaceutical companies to accomplish these objectives. (1) The DARC Communications Modules to integrate the DARC Structure Management System (SMS) and Reaction Management System (RMS) with any DBMS which provides host language interface facilities. These modules allow cross search, display of DBMS data in DARC and display of DARC structures in any DBMS session. (2) DARC-LINK 1 which allows the transfer of structures or queries between PC and mainframe. (3) DARC-MAC 1 which creates a metafile for use with Macintosh drawing software.

A chemical information system, in a chemical or pharmaceutical company requires access to 2 different types of data: the chemical structures handled by a chemical data base system and the information related to chemical structures including biological activity and toxicology handled by a data base management system (DBMS).

DBMSs are more generally perfectly serving their purpose except when chemical structures need to be inserted in screen or paper reports.

Chemical data base systems, and especially the DARC Structure Management System (DARC-SMS), have from the beginning offered sophisticated capabilities for chemical structure registration and structure, or substructure, or generic substructure search.

Nevertheless, research chemists also needed to have access from the structural system to the information such as biological test or inventory information stored in the DBMS; and biologists needed to have access from the DBMS to the structures stored in the chemical data base system.

In addition to that, it appeared that, despite the fact that relational DBMSs (RDBMSs) were offering a high level of performance, textual data were better handled by specialized text retrieval systems such as BASIS from Information Dimensions. The same need to combine chemical structures and data applied to text retrieval systems.

The chemical system and the DBMS or the text retrieval system had to communicate together. To solve this problem Télésystèmes has developed the DARC Communication Modules. The first version was released in 1986 and the software has been continuously improved from that date.

Also, the use of PCs, mainly IBM PCs or compatible PCs, but more and more often the Macintosh, is growing in the chemical and pharmaceutical industries.

A variety of software is offered on PCs and Macintosh to create local data bases and to offer local chemical reporting facilities for the research chemists.

The transfer of data from PCs to the central chemical data base or from the central chemical data base to PCs is then a critical point of the overall chemical information system.

To solve this problem Télésystèmes has designed in 1987 a DARC-F1 transfer format for both PCs and mainframe and also in 1987/1988, in cooperation with Janssen Pharmaceutica of Beerse, Belgium has developed an interface between DARC-SMS and the ChemDraw software produced by Cambridge Scientific Computing. During 1988 Télésystèmes was developing PC/host software named DARC-LINK 1.

DARC Communication Modules

Development Criteria. The development of the DARC Communication Modules took into consideration criteria which were considered as essential to build a chemical information system which meets the requirements of the researchers, while being easy to set up and to maintain.

The first criterion considered had to be a user-oriented access to both chemical structures and related data, which means for the researchers doing most of their work with the DMBS, an access to the chemical structures from the DBMS, and for the research chemist doing most of his or her work with DARC-SMS, an easy access to the related data, stored in the DBMS, from DARC-SMS.

Based on that first criterion, the DARC Communication Modules have been developed in such a way that they provide DARC-SMS with an easy and quick access to the DBMS data, but also easy and quick access to the chemical structures stored in DARC-SMS, from the DBMS, as well as easy cross search, starting either from DARC-SMS or from the DBMS.

The second criterion considered was related to access to the DBMS search capabilities. Since DBMSs have proven their ability to fulfill the needs of researchers in the chemical and pharmaceutical industries as well as in other fields of activity, it appeared to be essential that for cross searches, the DARC Communication Modules give full access to a DBMS session, or limited access depending on the needs or the allowance of the users.

The third main criterion considered was obviously related to high performance in communication between DARC-SMS and the DBMS.

To meet that requirement the DARC Communication Modules are based on a communication protocol (COM PROT in Figures 1–3) which takes advantage of all the capabilities of the VMS operating systems of Digital Equipment Corporation

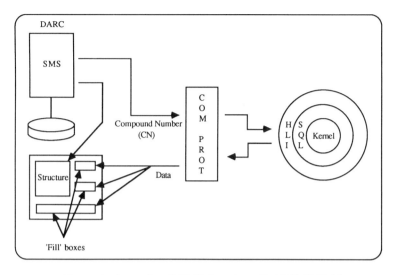

Figure 1. Accessing DBMS data from DARC-SMS (*1*).

(DEC) and is based on real time data transfer and parallel processing as illustrated in Figure 4.

Functionality.

Table I. The DARC Communication Modules

Function	Main Process	Communication Module
Display or print of DBMS data in DARC-SMS	DARC-SMS	FILL
Display of DARC-SMS chemical structures in the DBMS application (Structure Window)	DBMS	DISPLAY
Display of DARC-SMS chemical structures and DBMS data in the DBMS application using the full DARC-SMS display	DBMS	DISLAY CN
Cross search DARC-SMS and DBMS in both ways	DARC-SMS	DISPLAY/SELECT
Registration of chemical structures and DBMS data in a single application	DARC-SMS	INPUT CN

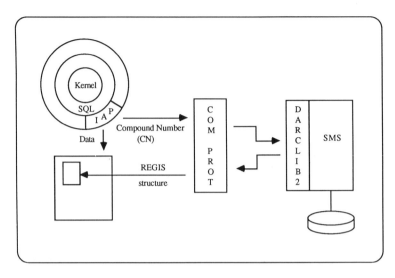

Figure 2. Accessing chemical structures stored in DARC-SMS from the DBMS (*1*).

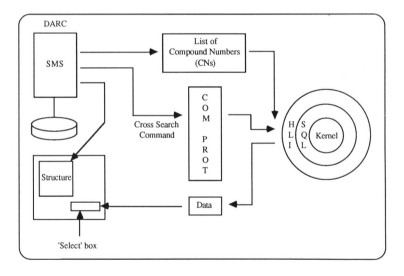

Figure 3. Cross searching DARC-SMS/DBMS.

Display or Print of DBMS Data in DARC-SMS: the "FILL" Module. The purpose of the "FILL" module is to allow the display or print of DBMS data in screen formats (see Figures 5–6) and the print of DBMS data in paper formats (A4) on PostScript laser printers (see Figure 7), both being defined using the DARC-RDS2 (Report Definition System 2) module (see Figure 8). The set-up of the "FILL" module is performed through the customization of the FILL Communication Protocol.

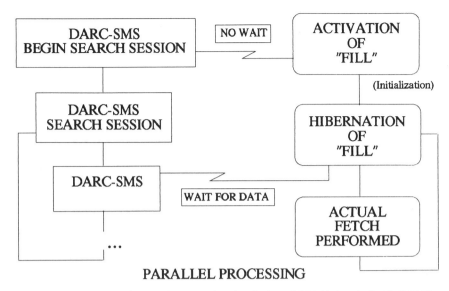

Figure 4. Real time and parallel process for the display of DBMS data in DARC-SMS.

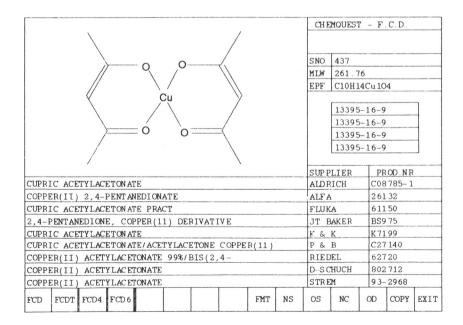

Figure 5. Display of DBMS data in a DARC-SMS user defined screen format.

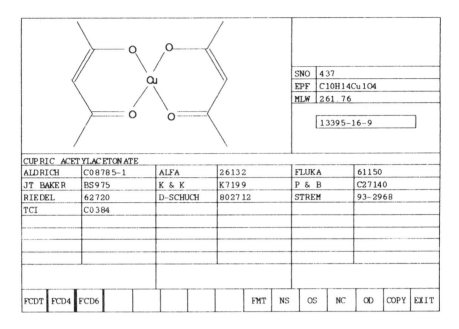

SNO	437
EPF	C10H14Cu1O4
MLW	261.76

13395-16-9

CUPRIC ACETYLACETONATE

ALDRICH	C08785-1	ALFA	26132	FLUKA	61150
JT BAKER	BS975	K & K	K7199	P & B	C27140
RIEDEL	62720	D-SCHUCH	802712	STREM	93-2968
TCI	C0384				

| FCDT | FCD4 | FCD6 | | | | | FMT | NS | OS | NC | OD | COPY | EXIT |

Figure 6. Display of DBMS data in a DARC-SMS user defined screen format.

The purpose of the set-up is to give access from DARC-SMS to all the information contained in the DBMS tables, and this set-up is performed by the data base manager. From that point the next step is the definition of the screen formats and printer formats using the user-oriented DARC-RDS2 module.

The definition is performed using a mouse for drawing boxes while specifying in each box which data from the DBMS will be displayed or printed in that box.

The printing of screen formats or paper formats (A4) on the PostScript laser printer is performed through the DARC-RPS (Report Printing System) module.

Display of Chemical Structures in the DBMS: the "DISPLAY" and "DISPLAY CN" Modules.

DISPLAY Module. The DISPLAY module can be called from any DBMS application to display a chemical structure in a graphics window, according to the needs of that application without any access to DARC-SMS. Indeed the DISPLAY communication protocol gives access to the DARC-SMS data base and performs the chemical structure display in the appropriate window (see Figure 9).

The display module works together with Tektronix and ReGis graphics.

The ReGis version is mainly used with ORACLE, the Tektronix version is used with other DBMSs such as System 1032 or RDB. A PostScript version is also available for printing chemical structures in DBMS reporting software using PostScript.

Substanzkarte

PRAEKLINISCHE-FORSCHUNG	Datum 15-09-88	Nr: SMS123-456

Von: W. Hofmann   AW: K-5524

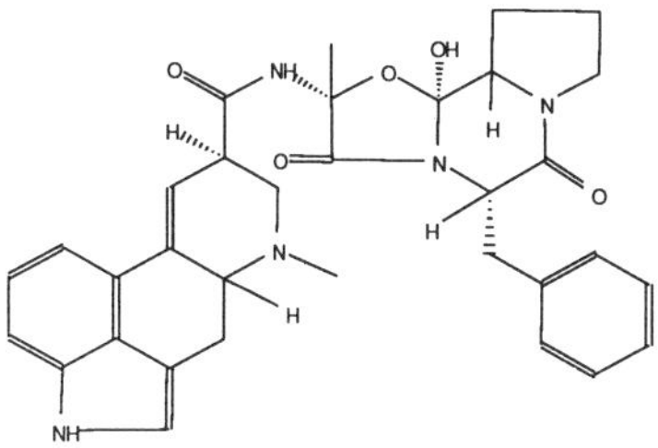

	BF:	C33H35N5O5
	MG:	581.68
	Smp:	212
	pKa:	

Anal.	PRF 4885
IR:	567833-A
NMR:	343499-V
MS:	134989-X

```
*** Ergotamine *** 12-Hydroxy-2-methyl-
  5-(phenylmethyl)ergotaman-3,6,18-trione
```

Bemerkungen: Vasoconstrictor
                specific in migraine

Lit.:   Stoll, Helv.Chim.Acta 46,2306 (1963)

Zu pruefen auf:

Syntheseweg:

Figure 7. Printing of DBMS data in a DARC-SMS user-defined printer format.

```
NAME : FCD  FIRST : FCD    NEXT : FCDT    CLASS :  0    NB BOXES : 42
NUMBER OF THE BOX TO DETAIL ?  55
                                              40 (CM)

                                              37 | 2  (SD)
                                              38 | 3  (SD)
                                              27 | 4  (SD)

                                                    21FILL A5
                                                    26FILL A5
                                                    41FILL A5
                                                    42FILL A5

1 (ST)                                        35 (CM)    | 36(CM)
5 FILL A0                                     6  FILL A1 | 7  FILL A2
8 FILL A0                                     9  FILL A1 | 10FILL A2
11FILL A0                                     12 FILL A1 | 13FILL A2
14FILL A0                                     15 FILL A1 | 16FILL A2
17FILL A0                                     18 FILL A1 | 19FILL A2
20FILL A0                                     22 FILL A1 | 23FILL A2
24FILL A0                                     25 FILL A1 | 28FILL A2
29FILL A0                                     30 FILL A1 | 31FILL A2
32FILL A0                                     33 FILL A1 | 34FILL A2
MEMORY ZONE :  DARC$MX 1        FIELD :  A0  OCCUR :   1
ANS :   1             TRUNC : Y  OUTLINE : Y  WORD : N  JUST : L
```

Figure 8. The definition of a screen format using DARC-RDS2.

DISPLAY CN Module. The DISPLAY module is very useful for all researchers having the need to display and print chemical structures in a DBMS application, without the need to have access to a DARC-SMS session.

The DISPLAY CN module brings extra capabilities based on the reporting facilities of DARC-RDS2, giving access from the DBMS to the complete display capabilities of DARC-SMS including the FILL communication module.

Thus DISPLAY CN allows the user to have, within a DBMS application, the same screen display which is obtained in DARC-SMS, including flexible report definition, display of 1 to 6 structures per screen, the use of graphics buttons for format selection and creation of PostScript files for printing on laser printers (see Figure 10).

Cross Search of Chemical Structures and DBMS Data: the "DISPLAY/SELECT" module. The purpose of the "DISPLAY/SELECT" module is to allow cross searching between DARC-SMS and the DBMS, starting either from DARC-SMS or from the DBMS and combining both types of cross searches (see Figure 11).

The set-up of the "DISPLAY/SELECT" module is performed through the customization of the DISPLAY/SELECT protocol.

The purpose of the set-up is:

1. To transfer a list of answers from DARC-SMS to the DBMS application, while calling from DARC-SMS the DBMS application. The DARC-SMS session is kept active while the DBMS application is running.

C	H	N	O	Cl	I	F	Br	S	Hal	Symbol	#Atoms
17		5	>5								

Acc.No.	Supplier	Prod.No.	CAS No.	Empirical Formula
Name				

22715	FLUKA	15389	25024-53-7	C17H19N5O8

BOC-N(IM)-DINITROPHENYL-L-HISTIDINE

26325	ALDRICH	86221-5		C17H17N5O7S

S-(2-HYDROXY-5-NITROBENZYL)-6-THIOINOSINE

26325	SIGMA	H8260		C17H17N5O7S

6-(2-HYDROXY-5-NITROBENZYL)THIOINOSINE

26326	ALDRICH	86149-9	38048-32-7	C17H17N5O6S

S-(P-NITROBENZYL)-6-THIOINOSINE

26326	VEGA	129881		C17H17N5O6S

6-P-NITROBENZYLTHIOINOSINE

Char Mode: Replace   Page 2   Count: 8

Figure 9A. A DBMS search application.

Acc. No. 26326	Molw. 419.42	Supplier	Prod. No.	CAS-No.
Empirical formula				
C17H17N5O6S				

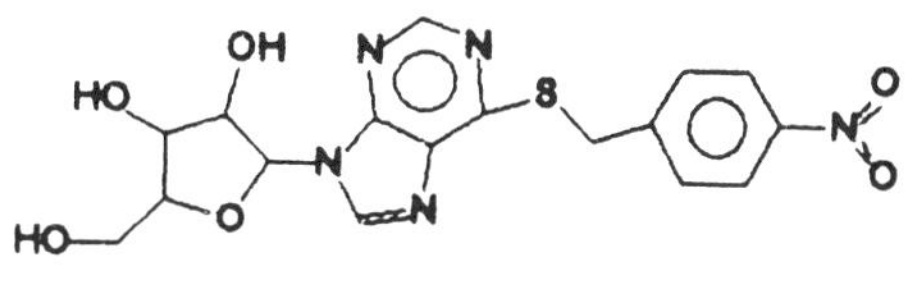

Supplier	Prod. No.	CAS-No.
ALDRICH	86149-9	38048-32-7
K & K	K4649	38048-32-7
SIGMA	N2255	
VEGA	129881	

S-(P-NITROBENZYL)-6-THIOINOSINE

Char Mode: Replace   Page 1   Count: *4

Figure 9B. The display of answers including the chemical structure.

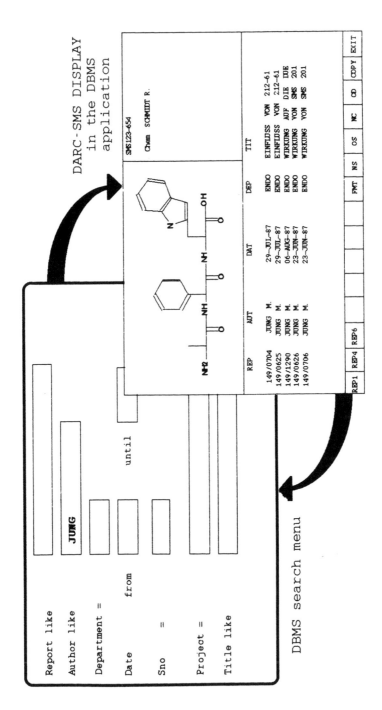

Figure 10. Calling the DISPLAY CN module from a DBMS application.

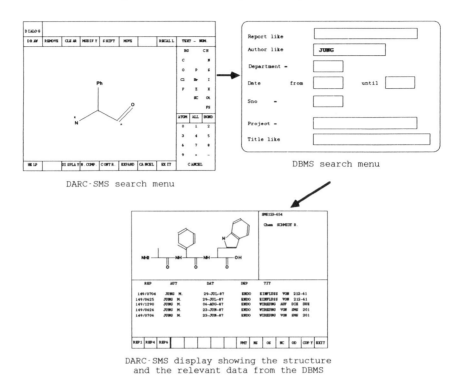

DARC-SMS search menu

DBMS search menu

DARC-SMS display showing the structure
and the relevant data from the DBMS

Figure 11. Cross search between DARC-SMS and a DBMS application.

2. To transfer a list of answers from the DBMS application to DARC-SMS when the DBMS session is completed, and directly to activate the DARC-SMS display within the DARC-SMS session which has remained active.

Besides the usual cross search, which allows the user to refine or expand a substructure search through a data search and again to refine the results of the combined search through a second substructure search, the DISPLAY/SELECT module offers an "offspring search" capability. Indeed the results of the initial substructure search within DARC-SMS are kept even after a data search in the DBMS, so that the researcher can always refine the results of his or her initial (sub)structure search through various data searched in the DBMS.

Cross-registration of Chemical Structures and DBMS Data: the "INPUT CN" module. The FILL, DISPLAY, DISPLAY CN and DISPLAY/SELECT Communication Modules combined together meet most of the requirements of the researchers for cross display of structures and DBMS data in the most appropriate way and for cross search. Nevertheless, despite the fact that in many cases the DBMS data registration is disconnected from the chemical structure registration, a need appeared for cross registration and data transfer between DARC-SMS and the DBMS to solve integrity problems.

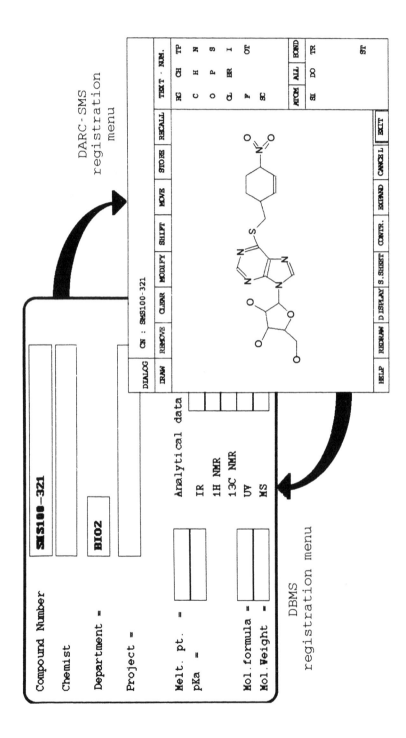

Figure 12. Chemical structure/DBMS data cross registration in a DBMS application.

The purpose of "INPUT CN" is to allow the registration of chemical structures in a DARC-SMS data base and the registration of data in a DBMS data base in the same application (see Figure 12). The assignation of the CN (Compound Identification) is performed within the DBMS application with visual control in the DARC-SMS registration menu and the molecular formula and molecular weight computed in DARC-SMS are transferred to the DBMS.

When using the INPUT CN communication module the DARC-SMS registration menu appears like a submenu of the DBMS application.

The actual registration of chemical structures, after the novelty check performed by DARC-SMS, can be performed within DARC-SMS or under the control of the DBMS application.

Communication with IBM PCs and Macintoshes

Chemical software on PCs has become quite popular during the last few years.

These PC software packages offer flexible solutions to research chemists for handling locally chemical structures.

The centralization of the chemical structures coming from a large environment of PCs, to build a corporate data base is not a simple problem and in any case requires quality control before a structure becomes validated in the corporate data base.

To help solve that problem, Télésystèmes has developed the DARC-LINK 1 solution in 1988.

DARC-LINK 1 is based on the DARC-F1 format.

The DARC-F1 format was initially developed for Markush-DARC to allow the transfer of Markush structures to the host from the PCs used by the data base producers with the Markush registration software.

DARC-F1 can easily describe structures, substructures or generic substructures and, of course, Markush structures.

The main functions of DARC-LINK 1, which is parametered according to the needs of the installation, are as follows:

1. Local conversion on the PC of chemical structures from any PC format to the DARC-F1 format (calling the appropriate conversion program).
2. Transfer of chemical structures written in the DARC-F1 format from the PC to the host.
3. Transfer of chemical structures written in the DARC-F1 format from the host to the PC.
4. Local conversion of chemical structures written in the DARC-F1 format to any format (calling the appropriate conversion program).
5. Loading of a Tektronix emulation program such as EMU-TEK Five Plus.

To allow quality control before integration of the transferred structures in the corporate data base, Télésystèmes has developed the concept of pseudo data bases (see Figure 13). A pseudo data base consists of all the chemical structures transferred by a research chemist from his or her PC to the host.

The data base manager can then recall these structures interactively for visual quality control and interactive novelty check.

DARC-LINK 1 also allows the transfer to PCs of chemical structures stored on

DRAWING CHECK

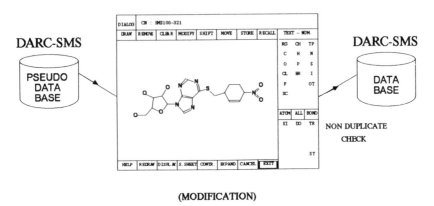

Figure 13. The concept of pseudo data bases.

the host in the DARC-F1 format, since DARC-SMS allows you easily to write the chemical structures resulting from any type of search in the DARC-F1 format.

Apple Computer Corporation's Macintosh probably offers the most flexible solution for desktop publishing thanks to its clipboard concept allowing you to copy and paste any chemical structure written in the appropriate format in any word processing software. In particular, Cambridge Scientific Computing's ChemDraw offers very high quality of drawings for chemical reports and many high quality word processors are available for the Macintosh.

Télésystèmes in cooperation with Janssen Pharmaceutica of Beerse, Belgium, has developed an interface which allows a user to write any chemical structure registered in a DARC-SMS database, in the ChemDraw format (see Figure 14).

The interface is based on a program allowing the user to convert chemical structures represented by DARC connection tables, into drawings such as those used by ChemDraw.

The ChemDraw compatible files can be created either from DARC-SMS or from external applications (DBMS applications) using libraries for conversion.

As far as the transfer from the host to the Macintosh is concerned, the current Apple-Talk connection program together with the appropriate software allows the users to access the files written on the host in the same way they access the files written on the Macintosh's hard disk. All these features, combined together, make the Macintosh one of the best solutions for the chemical reporting needs of the chemical and pharmaceutical industries.

Operating System, Hardware and Software

The DARC Communication Modules have been especially designed for the VAX and its VMS operating system, which have become the *de facto* standard for hardware and operating systems for the chemical and pharmaceutical industries.

MOTILIUM

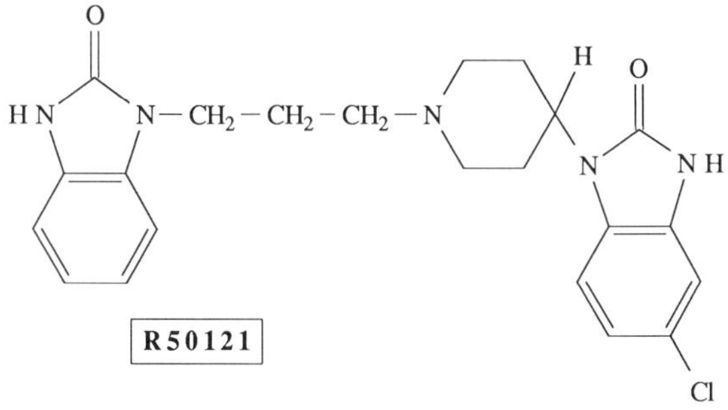

O
‖
H N —— N—CH$_2$—CH$_2$—CH$_2$—N

H   O
        ‖
    N   N H

R50121

Cl

R$_1$—N

H   O
        ‖
    N   N H

Cl

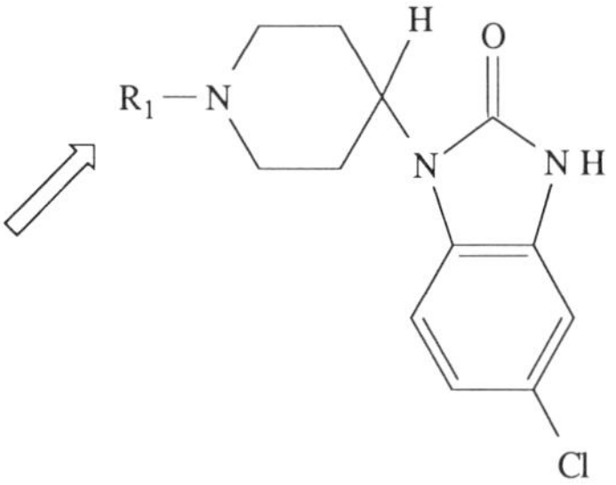

R$_1$			Activities		
			#1	#2	Global
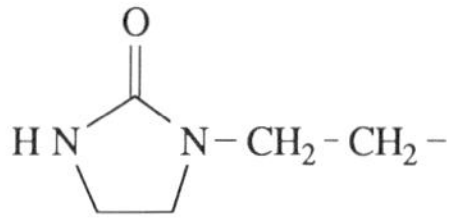 H N N—CH$_2$-CH$_2$– (O)	1.356	-0.74	A	C	
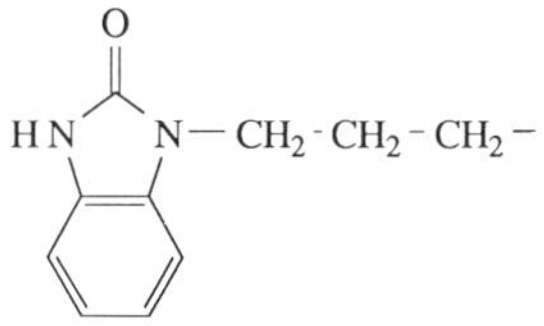 H N N—CH$_2$-CH$_2$-CH$_2$– (O)	1.246	-1.36	A$^+$	C$^-$	
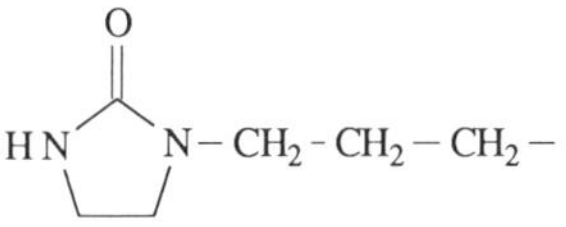 H N N—CH$_2$-CH$_2$—CH$_2$– (O)					

Janssen Pharmaceutica N.V.
1987

Figure 14. A chemical report produced on Macintosh using ChemDraw.

The first release of the DARC Communication Modules in 1987 was dedicated to Oracle Corporation's ORACLE RDBMS; the 2nd release was dedicated to Compuserve Software House's System 1032.

Versions compatible with DEC's RDB and with BASIS have been installed in 1988.

The use with other DBMSs or text retrieval systems such as Research Technology's INGRES or Paralog's TRIP is being investigated in cooperation with DARC-RMS users.

The DARC Communication Modules currently installed with VMS 4.7 have been tested by Télésystèmes with VMS 5 and only minor modifications have been required. Future releases of the DARC Communications Modules will take better advantage of the multiprocessor hardware environment of the new VAX series. As the current design of these modules is already based on parallel processing, the new VAX/VMS design appears to fit perfectly with the DARC Communication Modules concept.

Acknowledgments

The authors of this report gratefully acknowledge the contribution of the other members of the Télésystèmes development team and thank all the people in the chemical and pharmaceutical industries who contributed to the initial design as well as to the improvements of this chemical software development.

Literature Cited

1. de Jong, A.J.C.M. In *Chemical Structures: The International Language of of Chemistry*; Warr, W.A., Ed.; Springer Verlag: Heidelberg, 1988; pp 45–51.

RECEIVED May 2, 1989

Chapter 11

The Standard Molecular Data Format (SMD Format) as an Integration Tool in Computer Chemistry

H. Bebak[1], C. Buse[2], W.T. Donner[1], P. Hoever[1], H. Jacob[3], H. Klaus[1], J. Pesch[1], J. Roemelt[1], P. Schilling[4], B. Woost[1], C. Zirz[1]

[1] Bayer AG, D-5090 Leverkusen, West Germany   [2] Sandoz, CH-4002 Basel, Switzerland
[3] CibaGeigy, CH-4002 Basel, Switzerland   [4] BASF, D-6700 Ludwigshafen, West Germany

The Standard Molecular Data (SMD) format is described providing a powerful tool for data exchange between chemically oriented programs. By its modular design it offers a broad application range and great flexibility with respect to future extensions; i.e., definition of new information blocks is possible without affecting existing blocks. Furthermore, the SMD format includes the concept of superatoms, representing subsets of a molecular structure or ensemble, which offers a new strategy for economic and flexible storage of large molecules (proteins, polymers etc.) as well as permitting a new and consistent way for storage of reactions. This paper does not intend to give a technical report on the format, but the basic ideas and strategies guiding its design are presented.

Integration of chemistry programs is one of the major goals in computer chemistry today (1,2). (See also Donner, W.T. *Computational Chemistry in Industrial Research*, paper presented at the 7th International Conference on Computers in Chemical Research and Education, held in Garmisch Partenkirchen in 1985.) The reason is obvious as by no means all aspects of this field are covered by a single program system and the needs are still growing. Just to mention a few demands:

1. Retrieval of molecular structures including related data.
2. Reaction retrieval and synthesis planning.
3. Molecular modeling and physico-chemical parameters for the purpose of molecular design.
4. Molecular structure elucidation.
5. Quantum chemistry calculations.
6. Mixing of structures, reactions, data and text in reports.
7. Transport of commercial data bases into in-house systems.

There are programs available for each of these demands separately. But in general a project requires joint application of more than one task mentioned above

NOTE: This chapter is reprinted from *J. Chem. Inf. Comput. Sci.* **1989**, *29*, 1–5.

and for this purpose the efficient exchange of information between different systems is essential.

One way to overcome this problem is to equip the different systems with conversion routines, which enables the user to transfer data from one program to another (3). Here, one needs $N*(N-1)$ conversion routines for a set of N communicating programs. For a large number of programs it might be more advantageous to use a common data structure with a uniform interface. In this case the number of conversion routines does not exceed the number of programs. And for new program components to be integrated into the system it is just necessary to incorporate library routines which handle the input and output operations *via* the uniform data structure.

The Standard Molecular Data (SMD) format is designed to provide such an integration tool on the basis of a file format. It has been developed in the course of the CASP project (Computer Assisted Synthesis Planning) which is run by a consortium of seven German and Swiss Chemical Companies (BASF, Bayer, Ciba-Geigy, Hoechst, E Merck, Hoffmann La Roche and Sandoz). The basis of this development was the Molfile format of the earlier SECS program (4) (Simulation and Evaluation of Chemical Synthesis).

Recently some other research institutions in the field of computer chemistry exhibited distinct interest in the SMD format (e.g., Fraser Williams, ORAC Ltd, FIZ Chemie Berlin, Sadtler and others) and compatible formats have been designed for special purposes. For example there is the format by the Joint Committee on Atomic and Molecular Physics, JCAMP, for spectroscopic data storage which on one hand has a more limited scope, but on the other hand puts much emphasis on spectroscopically relevant features like stereochemistry. (Gasteiger, J.; Hendriks, B.M.P.; Hoever, P.; Jochum, J.; Somberg, H. "JCAMP-CS. A Standard Exchange Format for Chemical Structure Information in Computer Readable Form", to be published).

There is a variety of formats of similar scope in use, e.g., by CAS (5), DARC, MDL and others (6–8). But these support particular needs only, are restricted to use within special systems and hence, although being successful in their special fields, tend to be limited with respect to their information content and application range.

By no means do we claim that the SMD format to be presented here will cover all information explicitly that will ever be used in computer chemistry. But it is designed in an entirely modular form thus permitting definition and addition of completely new information without affecting the existing structures. This unique feature provides an extreme flexibility which ensures a nearly unlimited upward compatibility with respect to future developments.

While in the next section the general strategy and fundamental properties of the SMD format are presented, the general description is brought to life by discussing some selected examples in more detail in the following section. A few statements on the SMD library and the conclusions will complete the paper.

General Strategy and Fundamental Properties of the SMD

An SMD file is a sequential text file containing ASCII code. This ensures that the file is transferrable to different hardware irrespective of machine type (mainframes, departmental, workstations, PCs, etc.), its origin (IBM, DEC, etc.) and the programming language used. Furthermore, this allows you to read and to interpret the file visually, which sometimes is convenient, and to edit its contents with any text editor. However, the latter practice is not recommended as it can easily lead to files which are no longer readable by the SMD routines themselves and which might be inconsistent in their information.

The information content of an SMD file is organized in an hierachical order with the following structural tools:

SMD File. The SMD file represents the global connection to the operating system. It may contain a single SMD structure or several SMD structures.

SMD Structure. This entity includes the information with respect to a chemically relevant unit, i.e., a single molecule or an ensemble of molecules which are in a certain relation to each other (e.g., a reaction). It is our strategy that such an SMD structure is complete and consistent in its information content. There are no pointers to external items, which might be modified in an uncontrolled way and thus spoil the overall relation. For reactions this requires the storage of reactants and products for each reaction, again, instead of establishing connections to the corresponding molecular items. This procedure seems to be a very inconvenient way of storage. But on the other hand this strategy dispenses us from setting up a global checking system, keeping track of the interrelations between separate information units. In our opinion such a checking system is not manageable in large data bases of molecular structures and related reactions. Furthermore, as soon as one tries to store entire reaction sequences or even reaction networks with the associated atom correspondencies, one will be lost with the idea of separate information units for reactants and products.

Information Blocks. Each SMD structure is divided, again, into several information blocks each containing special data associated with the corresponding SMD structure. These blocks are to serve as simple transport mediums for information, i.e., no checking on its consistency and correctness is done. It is entirely up to the writing program to ensure the validity of the information and the reading system to interpret only those blocks which it is able to handle.

A minimal list of information blocks needed for storage of molecular and reaction data is given in Table I. Any extension of this set is very easy. In effect, the list of information blocks actually used will be considerably larger and entirely fitted to the needs of the programs within a system. Even graphics data, represented by some kind of graphics metafiles, may be included and transferred via the SMD format. Its modularity makes it extremely flexible, open to any additional data and new data types can be supplemented without affecting already existing blocks.

Subblocks and Superatoms. In order to allow for additional structuring of the information in a block the subblock tool is provided. Besides the general strategy of ordering information this idea leads to the concept of superatoms with respect to the storage of chemical structures (i.e., structuring the CT, CO and the LB blocks). Hence, superatoms represent molecular fragments of a large structure or molecular

Table 1.   Minimal list of information blocks needed for storage of molecular and reaction data

block type	block content
DTCR	creation date and time of the structure (cf. fig. 2)
DTUP	date and time of last modification (cf. fig. 2)
CT	connection table(s) of a structure or a reaction (cf. figs 2–4)
CO	corresponding cartesian coordinates (cf. fig. 2)
LB	atom/superatom labels (any descriptors or labels attached to an atom or superatom) (cf. figs. 2–4)
FORM	empirical formula (cf. fig. 2)
NAME	name of compound (any name of compound e.g. trade name, IUPAC name etc.; not standardized) (cf. fig. 2)
TEXT	arbitrary text information related to the molecule or reaction (cf. figs. 2, 4)
DDS	general data related to a single compound (dynamical data set block; i.e. CAS-Reg. No., special properties, test results with respect to activities etc. may be stored here; cf. fig. 2)
RXN	general data related to a reaction (reaction data block; cf. fig 4)

entities in the frame of a complete reaction. In general, a superatom can be any arbitrary subset of the chemical unit represented in the SMD structure. This concept of an hierarchical structured representation of chemical structures permits completely new strategies in the storage of molecules and reactions. A more detailed discussion with some examples demonstrating this feature is given in the next section.

Data Record. Data records are the lowest level in the information hierarchy and provide the explicit information. They have to start with a letter, a digit or a space. This requirement implicitly defines all other characters as potential control characters. If the remaining columns at the end of a record have the values zero or space they can be omitted. All data within a record are kept in free format unless a format is specified in the block or subblock itself.

In summary, there is a hierarchy of tools to structure the information in an SMD file. A schematic organization of an SMD file is represented in Figure 1. It is completely up to the communicating programs to use these tools for an intelligent, convenient and economic transfer of data.

Discussion of Examples

After having presented the basic ideas behind the design of the SMD format the following discussion of some selected examples is meant to demonstrate the practical use of the tools described above.

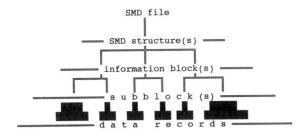

Figure 1. General scheme of information levels in a SMD file.

Example 1: SMD Structure of 2-Hydroxypyridine-N-oxide. Our first example deals with a single molecule. The molecular diagram and the corresponding SMD file are given in Figure 2 with detailed information on the meaning of the different entries described in italic letters. The SMD file starts with a special start record for an SMD structure:

> STRT {name}

with {name} being an optional classification of the corresponding SMD structure. It is finished implicitly either by the starting record of the next SMD structure within the file or by the end of file mark (as in our case). A series of information blocks (DCTR, DTUP, CT, CO, LB, FORM, NAME, TEXT and DDS) follows, each identified by a start record of the form:

> {blktyp} {name} {blklen} {text}

The abbreviations used stand for
>   special delimiter indicating the block level of information hierarchy
{blktyp}   type of information block (up to four characters)
{name}   an arbitrary optional classification
{blklen}   number of data records within this block excluding the starting record;
           (5 digits, optional, right justified)
{text}   optional text

Each block is finished either by the next delimiter ">", the delimiter of a subblock "]" (see below), or the end of file mark. The contents of the different blocks are described in Figure 2. A few of the blocks require some additional comments:

CT block: To represent chemical structure information in computer readable forms, many methods have been proposed (*9–14*). According to our general policy mentioned in the previous section we prefer a topological representation of structures in the form of a connection table (CT) rather than a linear notation. We

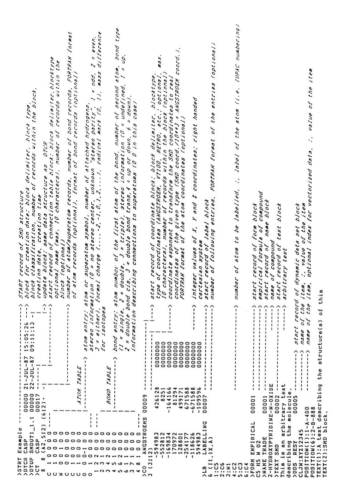

Figure 2. SMD structure of 2-Hydroxypyridine-N-oxide.

use a noncanonical and nonconcise connection table. If necessary, the program that reads the SMD file could transform it into the canonical representation by performing rules originally proposed by Morgan (*15*) and improved by Moreau (*16*). The CT includes the basic data inherent in structural diagrams such as atom properties (elements of the periodic table, superatoms, charges, radical and stereo information) and bond properties (connectivities: single, double, triple; stereo information). With the exception of some stereochemical information, the SMD format does not consider artificial atom or bond types (aromatic, tautomeric, ring etc.), which are normally the result of a perception process. But it is not difficult to make provision of such atom or bond types in future versions of the SMD format, because this requires definition of some additional atom or bond entries in the CT block only. However, at the moment we do not see a commonly agreed definition of this additional atom and bond information and therefore we restrict the SMD format to the conventional description.

CO Block: The coordinates are stored as integer values. The actual units used (angstrom, au, relative screen units, etc.) are specified in the starting record, while the exponent to convert the integers to the true floating point values is given in the first data record.

LB Block: In the LB block any arbitrary characterizing labels may be attached to the atoms (e.g., IUPAC numbering).

DDS Block: The Dynamical Data Set (DDS) block enables the storage of any data related to the SMD structure. The different items can be defined dynamically in scalorized as well as in vectorized form.

As our first example represents a single and rather small molecule, the corresponding information structure is quite simple (cf. Figure 3).

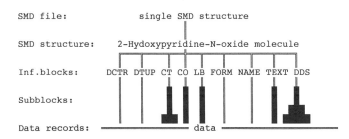

Figure 3. Information hierarchy in the SMD structure of 2-Hydroxypyridine-N-oxide given by Figure 2.

Example 2: SMD Structure of a Polyphenylsulfide-Polymer. In this example, the SMD format is used for an economic storage of a polymer by representing it as a chain of its monomers, with each monomer in the chain described as a superatom. As can be realized from Figure 4 the CT block is structured *via* a subblock CTP1.

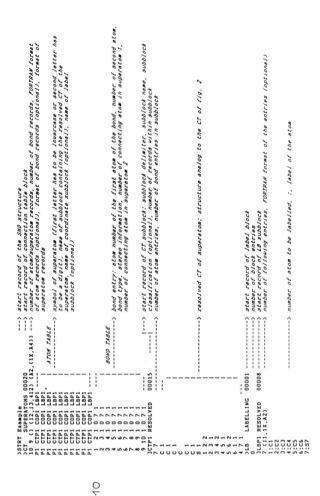

```
>STRT Example                        --> start record of the SMD structure
>CT  SUPERATOMS 00020                --> start record of connection table block
10 9 () (I2,I3,4I2) (A2,(1X,A4))     --> number of atom/superatom records, number of bond records, FORTRAN format
                                         of atom records (optional), format of bond records (optional), format of
                                         superatom records
P1 CTP1 COP1 LBP1,      ATOM TABLE   --> symbol of superatom (first letter has to be lowercase or second letter has
P1 CTP1 COP1 LBP1                        to be a digit), name of subblock containing the resolved CT of the
P1 CTPE COP1 LBP1                        superatom, name of coordinate subblock (optional), name of label
P1 CTP1 COP1 LBP1                        subblock (optional)
P1 CTP1 COP1 LBP1
P1 CTP1 COP1 LBP1
P1 CTP1 COP1 LBP1
P1 CTP1 COP1 LBP1
P1 CTP1 COP1 LBP1
P1 CTP1 COP1 LBP1
1  2 1 0 0 7 1         BOND TABLE    --> bond entry: atom number of the first atom of the bond, number of second atom,
2  3 1 0 0 7 1                           bond type, stereo information, number of connecting atom in superatom 1,
3  4 1 0 0 7 1                           number of connecting atom in superatom 2
5  6 1 0 0 7 1
6  7 1 0 0 7 1
7  8 1 0 0 7 1
8  9 1 0 0 7 1
9 10 1 0 0 7 1
JCTP1 RESOLVED 00015   |----|        --> start record of CT subblock: subblock delimiter, subblock name, subblock
7             00015     |----|           classification (optional), number of records within subblock, number
C 1 1                                    of atom entries, number of bond entries in subblock
C 1 1
C 1 1
S 1 2
1  2                                 --> resolved CT of superatom: structure analog to the CT of fig. 2
2  3
3  2
4  5
5  6
6  1
1  7
>LB  LABELLING  00001                --> start record of label block
0             00008                   --> number of block entries
JLBP1 RESOLVED 00008                 --> start record of LB subblock
7 (I1,1X,A2)                          --> number of following entries, FORTRAN format of the entries (optional)
1:C1
2:C2
3:C3                                 --> number of atom to be labelled, :, label of the atom
4:C4
5:C5
6:C6
7:S7
```

Figure 4. SMD structure of a Polyphenylsulfide polymer.

While the CT block represents the chain of monomers the explicit structure of the monomer is described only once in the CTP1 subblock. Hence an extremly compact storage of the polymer is achieved. Similar strategies of structured representation will be applicable for any large molecule (e.g., proteins etc.) and they provide an interesting tool with respect to quantitative structure activity relationship (QSAR) studies.

Technically, the routines automatically assume a superatom structure, because the atom identifications in columns 1–2 of the data record in the CT block do not specify an allowed atom type. In this case the following fields are interpreted as pointers to the corresponding CT, CO and LB subblocks within the SMD structure. (In our example only the CTP1 and LBP1 subblocks are given explicitly. But it is easy to realize that an analogous subblock structure might be present, if specified at all, for the coordinate and label block, too. Furthermore the reader should be aware of the fact that any CO subblock will contain relative coordinates shifted to an arbitrary origin of the molecular fragment.) In the case of such a superatom structure the last two entries of the bond specification list gain their full importance. They specify the connecting atoms in superatom 1 and/or 2, respectively, according to the numbering used in CT subblock CTP1.

Although for clarity of the example no information blocks other than CT and LB have been included, the information structure in this example is more sophisticated compared to example 1 (cf. information hierarchies in Figures 3 and 5).

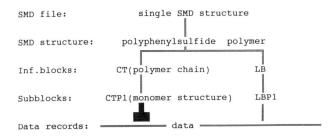

Figure 5. Information hierarchy in the SMD structure of the Polyphenylsulfide-polymer described by Figure 4.

Example 3: SMD Structure for a Reaction. This example is to demonstrate how the concept of superatoms can be used for storage of reactions. The reaction example and its corresponding SMD structure is given in Figure 6. There are two superatoms (ed and pr) representing the starting materials (educts) and product ensemble, respectively, which are not connected to each other. Possible intermediates might have been described by further independent superatoms. The educt subblock is split again *via* a nested subblock structure into the two reactant molecules. In an analogously organized but not fully expanded label block structure the different molecules are identified as reactant 1, reactant 2 and product.

In the RXN block including subblock structure as well as dynamic data set

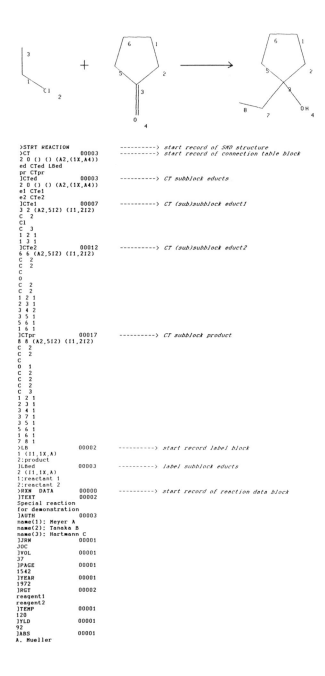

Figure 6. SMD structure of a reaction.

facilities, the corresponding reaction data (e.g., yield, temperature, references, compound information etc.) can be stored.

The information structure tree related to this reaction example is given by Figure 7.

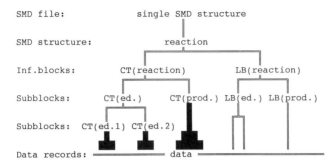

Figure 7. Information hierarchy in the SMD structure for the reaction described by Figure 6.

SMD Library

Once a standard file format has been designed, all relevant programs have to be interfaced to it. The amount of work for this step can be minimized by providing a library of subprograms which can be included in a program to handle the input/output operations. An idealized scheme of such a set of integrated programs is given in Figure 8.

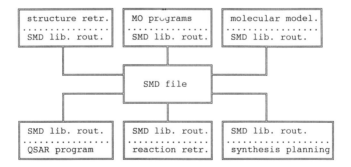

Figure 8. Scheme of an integrated program system.

The quality of such a common package is certainly better than the quality of several isolated routines since it is used by a large number of persons in quite different ways. Efficient validation programs can be provided which check on a formal basis, whether a given file conforms to SMD or not. Additions to the SMD format are included in the subprograms of the library; additions to a program are only necessary if the data involved in this addition are relevant. The most important point is the interference of the subroutines with the internal representation of the connection table within the program. The package has been designed in such a way that a programmer has to provide a few subroutines transferring data from or to the internal connection table. These calls remain unchanged even if the format itself and/or the library are changed.

The library was developed using FORTRAN under TOPS-20 and VMS, but every effort was made to adhere to standards so that the package should run on any machine with a FORTRAN compiler.

Conclusion and Perspectives

Comparing the different information trees of the examples 1–3 above it is realized that the SMD format described in this paper provides an adequate and flexible tool for transfer of chemical structure information and data between chemically oriented programs. Its modular design offers a broad application range and guarantees nearly unlimited upward compatibility to future extensions. And there is no doubt that certain points will require extension in the future, e.g., stereochemical description of molecules, tautomeric and mesomeric bonds, coordinate bonds in inorganic and metal organic complexes, storage of reaction sequences including atom correspondences, etc. Research on these and even more topics still goes on. But as soon as widely agreed solutions to one or several of these projects is achieved the SMD format is ready to be extended with respect to them. The value of the format as a transfer medium depends on simple and generally accepted conventions. Specific interpretations of chemical features should not be used within the SMD format. SMD library routines and, if necessary, special perception routines are to be included within the individual programs to obtain a particular interpretation from basic information of the SMD format (and *vice versa*).

The concept of superatoms included in the SMD format permits a structured representation of (large) molecules and reactions. Furthermore we believe that this concept indicates a possible way of handling generic structures and Markush formulas. We do not have a final solution for these problems yet, but promising attempts are under investigation.

Note

A detailed technical description of the SMD format and the corresponding SMD library will be provided upon request by:

Dr W T Donner
BAYER AG
ZF-DID, Geb. Q18
D-5090 Leverkusen –1–
Federal Republic of Germany

Acknowledgments

The SMD format presented here is the result of many fruitful discussions by the authors with W. Boell (BASF), H. Braun (Hoffmann La Roche), H. Bruns, L. Krakies (E. Merck), J. Sander, W. Schwier (Hoechst), W Sieber (Sandoz) and R. Wehrli (Ciba-Geigy) which had considerable impact on its design.

Literature Cited

1. Williams, M.; Franklin, G. In *Chemical Structures: The International Language of Chemistry*; Warr, W.A., Ed.; Springer Verlag: Heidelberg, 1988; pp 11–21.
2. Hagadone, T. In *Chemical Structures: The International Language of Chemistry*; Warr, W.A., Ed.; Springer Verlag: Heidelberg, 1988; pp 23–41.
3. Pensak, D.A. *Ind. Res. Dev.* **1983,** *25*, 74–78.
4. Wipke, W.T.; Dyott, T.M. *J. Am. Chem. Soc.* **1974,** *96*, 4825–4842.
5. *Chemical Abstracts Service Registry Structure Standard Distribution File*; Chemical Abstracts Service: Columbus, Ohio, 1977.
6. Bernstein, F.C.; Koetzle, T.F.; Williams, G.J.B.; Meyer, E.F.; Brice, M.D.; Rodgers, J.R.; Kennard, O.; Shimanouchi, T.; Tasumi, M. *J. Mol. Biol.* **1977,** *112*, 535–542.
7. Allen, F.H.; Bellard, S.; Brice, M.D.; Cartwright, B.A.; Doubleday, A.; Higgs, H.; Hummelink, T.; Hummelink-Peters, G.; Kennard, O.; Motherwell, W.D.S.; Rodgers, J.R.; Watson, D.G. *Acta Crystallogr.* **1979**, *B35*, 2331–9.
8. Crennell, K.M.; Brown, I.D. *J. Mol. Graphics* **1985,** *3*, 40–49.
9. Rush, J.E. *J. Chem. Inf. Comput. Sci.* **1976,** *16*, 202–210.
10. Dromey, R.G. *J. Chem. Inf. Comput. Sci.* **1979,** *19*, 37–42.
11. Nakayama, T.; Fujiwara, Y. *J. Chem. Inf. Comput. Sci.* **1980,** *20*, 23–28.
12. Barnard, J.M.; Lynch, M.F.; Welford, S.M. *J. Chem. Inf. Comput. Sci.* **1982,** *22*, 160–164.
13. Nakayama, T.; Fujiwara, Y. *J. Chem. Inf. Comput. Sci.* **1983,** *23*, 80–87.
14. Rayner, J.D. *J. Chem. Inf. Comput. Sci.* **1985,** *25*, 108–111.
15. Morgan, H.L. *J. Chem. Doc.* **1965,** *5*, 107–113.
16. Moreau, G. *Nouv. J. Chim.* **1980,** *4*, 17–22.

RECEIVED May 2, 1989

Chapter 12

The Effort To Define a Standard Molecular Description File Format

John S. Garavelli
**Biomolecular Analysis Lab, M/C 781, University of Illinois at Chicago,
Chicago, IL 60680**

In the past year some progress has been made toward achieving a standard molecular description file format. Proposals for a standard file format were advanced separately by two groups at the Gordon Conference on Computational Chemistry in July 1988. At the Workshop on Standards for Exchange of Computerized Chemical Structures conducted at the September 1988 American Chemical Society National Meeting in Los Angeles the work of a consortium of European chemical and pharmaceutical companies was presented and an international committee was organized to continue development of this standard and to seek its recognition from IUPAC and CODATA. In December 1988 a standard format was proposed for the publication of molecular modeling results in the medicinal chemistry area. The format proposed by the European consortium, the SMD proposal, seems to address the needs for modeling small molecules, for molecular graphics and for reactions, and it may be adequate for the needs of quantum chemistry. But the broader range of information which a molecular description file might be expected to convey imposes additional design requirements on a standard molecular description file format.

The Need for a Standard

Many people doing molecular modeling use more than one program to construct and display molecular structures and perform calculations based on those structures. Each molecular modeling program usually has a file format for storing molecular structures which is peculiar to itself and not generally usable by any other program. Confronted with the problem of transferring molecular structure data between molecular modeling programs, a molecular modeler may take one of three approaches. The first approach, the one initially undertaken by most modelers, is to write a series of programs each of which transforms a molecular model stored in one format to another format. This is illustrated in Figure 1. Taking this approach with N incompatible modeling programs, the modeler has the formidable task of writing and maintaining $N^2 - N$ different transformation programs. In addition new programs may have to be written for each new release of a software program. Some

Modeling Program	Conversion Programs			
AIMS/ECEPP	—	AIM2AMB	AIM2CHA	...
AMBER	AMB2AIM	—	AMB2CHA	...
CHARMM/HYDRA	CHA2AIM	CHA2AMB	—	...
CHEMLAB	CHE2AIM	CHE2AMB	CHE2CHA	...
DISCOVER/INSIGHT	DIS2AIM	DIS2AMB	DIS2CHA	...
MACROMODEL	MAC2AIM	MAC2AMB	MAC2CHA	...
MM2	MM22AIM	MM22AMB	MM22CHA	...
SYBYL/MENDYL	SYB2AIM	SYB2AMB	SYB2CHA	...
BROOKHAVEN	PDB2AIM	PDB2AMB	PDB2CHA	...
CAMBRIDGE	CAM2AIM	CAM2AMB	CAM2CHA	...
...	...	...	...	...

Figure 1. Writing a program for each possible transformation between N incompatible molecular description file formats requires $N^2 - N$ programs.

software companies are considerate enough to supply such conversion programs for different release versions of its own programs and occasionally a far-sighted company may even supply conversion programs for its format and other non-proprietary formats.

The second approach, illustrated in Figure 2, is to write one program with N subroutines each reading one type of format and transforming it into an internal representation, and N subroutines each using the internal representation to write

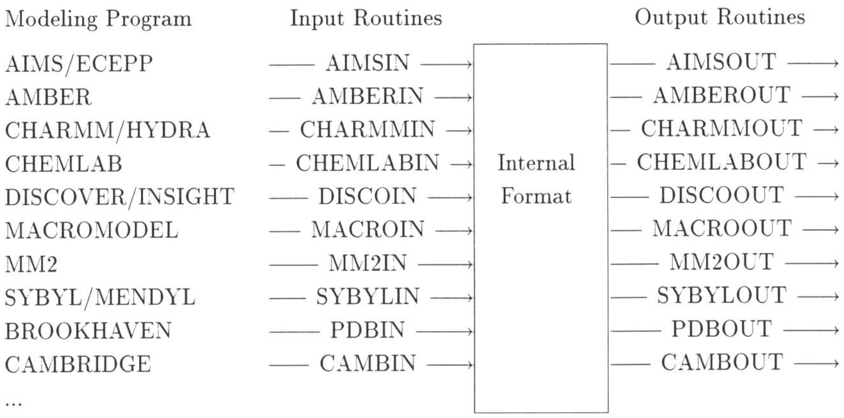

Modeling Program	Input Routines		Output Routines
AIMS/ECEPP	—— AIMSIN ——→		—— AIMSOUT ——→
AMBER	—— AMBERIN ——→		—— AMBEROUT ——→
CHARMM/HYDRA	— CHARMMIN →|		— CHARMMOUT →
CHEMLAB	– CHEMLABIN →|	Internal	– CHEMLABOUT →
DISCOVER/INSIGHT	—— DISCOIN ——→	Format	—— DISCOOUT ——→
MACROMODEL	—— MACROIN ——→		—— MACROOUT ——→
MM2	—— MM2IN ——→		—— MM2OUT ——→
SYBYL/MENDYL	—— SYBYLIN ——→		— SYBYLOUT ——→
BROOKHAVEN	—— PDBIN ——→		—— PDBOUT ——→
CAMBRIDGE	—— CAMBIN ——→		—— CAMBOUT ——→
...			

Figure 2. Writing a program with an internal representation and subroutines for input and output of N incompatible molecular description file formats requires 2N subroutines.

one type of format. Taking this approach, the modeler has to write and maintain only 2N subroutines each about as complex as one of the programs required by the first approach.

The third approach, and maybe the most difficult, is for molecular modelers to agree that all molecular modeling programs should read and write the same file format for storing molecular structures. Taking this approach, illustrated in Figure 3, an individual modeler would have to write and maintain 0 programs and subroutines, an obvious saving in time, effort and money over either the first or second alternatives. The difficult part is getting everyone concerned to agree on the standard. Commercial molecular modeling programs could offer such a standard file format as an option for storage of models. The creators of the N molecular modeling programs would have to be convinced that it is more cost effective to write, or replace, 2 subroutines in their own programs thereby making those programs more user-friendly, attractive and salable.

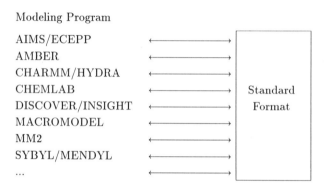

Figure 3. With a standard molecular description file format modelers would not have to write conversion programs.

One serious problem which is encountered in the conversion of molecular description files and which could be ameliorated by the adoption of a standard is the mapping of atom types from one program to another. Atom types in molecular modeling programs seem to be uniquely defined by atomic number, atomic mass, formal charge, hybridization state and the atom type of bonding partners. Unfortunately, even if all atoms, particularly hydrogen, are explicitly included in a molecular description along with formal charges, unpaired electrons and connection information, it may not be possible to assign a particular modeling program's atom types to some atoms in a model. The incomplete parameterization of modeling programs means that the mapping of atom types may be "on-to" but not "one-to-one".

History

There are two molecular description file formats which can be read by many public domain and commercial programs. The Cambridge Crystallographic Data Base

(CCD) distributed by the Crystallographic Data Center, University Chemical Laboratory, Lensfield Road, Cambridge, CB2 1EW, England and by the Medical Foundation of Buffalo, 73 High Street, Buffalo, NY 14203, USA, contains coordinates of small organic and inorganic compounds. The Brookhaven Protein Data Bank (PDB) contains coordinates of proteins and a few nucleic acids and polysaccharides. The PDB is distributed by the Chemistry Department, Brookhaven National Laboratory, Upton, NY 11973, USA. The CCD format does not lend itself to the representation of polymeric molecules, and the bonding or connectivity information is maintained in a separate file from the atomic position information. The PDB format assumes bonding within polymeric units but provides a means of conveying bonding information for heterologous units. Being used principally to represent X-ray data, both formats usually neglect the representation of hydrogen atoms and charge information.

The International Union of Crystallography adopted a Standard Crystallographic File Structure (SCFS) in 1984 (*1*) and revised it in 1987 (*2*). The SCFS format employs a block structure but is otherwise similar to the PDB format with provision made for valence, charge and bonding information. However, the SCFS format has not been adopted for use by any molecular modeling programs because neither the Cambridge Crystallographic Database nor the Brookhaven Protein Databank has adopted, and been distributed in, the SCFS format.

There are at least four file formats for biopolymer sequence data including the formats of the National Biomedical Research Foundation protein and nucleic acid sequence data bases (NBRF), the European Molecular Biology Laboratory Nucleotide Sequence Data Library (EMBL), the Genetic Sequence Data Bank (GenBank), and the New Atlas of Protein Sequences (NEWAT). (The Protein Identification Resource protein and nucleic acid sequence data bases are distributed by the National Biomedical Research Foundation, Georgetown University Medical Center, 3900 Reservoir Road, NW, Washington, DC 20007, USA. The EMBL Nucleotide Sequence Data Library is distributed by The European Molecular Biology Library, Postfach 102209, D-6900 Heidelberg, FRG. The GenBank is distributed by IntelliGenetics, Inc., 700 East El Camino Real, Mountain View, CA 94040, USA, in collaboration with the Los Alamos National Laboratory. The New Atlas of Protein Sequences was distributed by Dr. Russell F. Doolittle, Department of Chemistry, M-034, University of California at San Diego, La Jolla, CA 92093, USA.) In 1987 the Committee on Data for Science and Technology (CODATA) Task Group on Coordination of Protein Sequence Data Banks proposed a standard file format for protein sequence data (*3*). These formats represent only sequences of biopolymer units, chemical structures without atomic positions. However, as increasing numbers of protein structures must be derived from nucleic acid sequences and conformation prediction methods, the interconversion of biopolymer sequence data and "all atom" molecular descriptions will become increasingly important for molecular modelers.

About 1980 as an outgrowth of a project for computer assisted synthesis planning (CASP), a consortium of European chemical and pharmaceutical companies began working on a Standardized Molecular Data, SMD, file format for interchange of chemical information. Version 4.0 of the SMD format was produced in August 1986 and version 4.3 in February 1987. At the Workshop on Standards for Exchange of Computerized Chemical Structures conducted at the 196th American

Chemical Society National Meeting in Los Angeles in September 1988 the work of the SMD consortium was presented by Wolfgang T. Donner and John M. Barnard. The SMD format like SCFS employed a block structure but was otherwise similar to the CCD format. Some effort was made to employ non-alphanumeric symbols rather than keywords so that language bias could be minimized. Another objective in the formulation of the SMD format was the representation of chemical reactions along with chemical structures. Many participants in the September 1988 ACS Workshop felt that the proposed SMD format would be adequate for the needs of small molecule modeling, molecular graphics and reaction representation but that it might not meet the additional requirements of quantum chemistry, crystallography, molecular dynamics and macromolecular modeling. At that workshop an international committee was organized to continue development of the SMD standard and to seek recognition for the proposal from such organizations as IUPAC and CODATA.

Proposals for a standard file format were advanced separately by DeLos F. DeTar and by T. J. O'Donnell and John S. Garavelli at the Gordon Conference on Computational Chemistry in July 1988. Both these proposals were advanced to address problems confronting molecular modelers. The proposal of DeTar provided for representation of force field data as well as molecular structure information. O'Donnell and Garavelli presented several general formats including an alternative to fixed column, record oriented formats.

In December 1988 an article in the *Journal of Medicinal Chemistry* presented in its guidelines for publication a "General Molecular File Format" for publication of molecular modeling results in the medicinal chemistry area (*4*). This format employed fixed length records with fixed field positions and was similar to PDB format but with the inclusion of formal and partial charge, atom type, atom names and atom labels.

Development Considerations

A molecular description file might be called upon to represent chemical information in any of the following areas: quantum mechanics, crystallography, molecular diagrammatics and graphics, molecular mechanics and dynamics, structure comparison and substructure searching, property prediction, subunit sequence comparison, conformation prediction, synthetic pathway generation and reaction parameter estimation. Consideration of the broad range of that information and of approaches offered by advances in computer science and data base management suggests that a standard molecular description file format should meet the following requirements.

1. It must store information about molecular structure which would be essential to draw a complete, correct chemical structure. This information would include atomic number, bonds, formal charge and unpaired electron information. For some but not all applications, atomic coordinate and atomic mass number (isotope) information would also be essential.
2. It must include enough information for molecular mechanics programs to complete assignment of "atom types." This requirement, so far as is known, would be fulfilled by meeting the first requirement, but conceivably that may not be sufficient.

3. It must be versatile to accommodate additional information which is extrinsic to molecular structure, such as atomic charges, atomic velocities or temperature factors.

4. It must be extensible for local user adaptations and for broad developmental advances. But, the method for making extensions and their specifications must be provided in the standard. A file containing local extensions must contain the specification for those extensions in the standard format so that it can be read by any program which purports to read the standard. A program reading local extension material will recognize its construction, confirm its syntax and then may use it or ignore it.

5. The standard format specification should be written in Backus-Naur Form (BNF). The specification in BNF would go to the level of primitive object and construction of macro definitions and then be built up from those macro definitions. Updating of the standard would be accomplished by adding macro definitions. In this way programs conforming to the standard at a particular version level, should still be able to read a file written at any version level though they would not be able to use the information in later version constructs. This requirement should provide downward and upward compatibility as the standard is revised.

6. The standard format specification should be complemented by release of programs for standard verification.

The features which a standard molecular description file format must have in order to be useful and enduring largely depend on the research areas in which the file is to be used. But certainly, it would be helpful if the standard were versatile and extensible so that emerging areas of research could be accommodated without frequent major revisions being necessary. However, the scale of detail in the models employed in some research areas differs so greatly that it may be better to have different but related standards for each area. Specifically, perhaps there should be a standard for molecular descriptions at the atomic level and a separate standard for biopolymer descriptions at the residue level. The two standards must be related so that a molecular model at the atomic level could be generated for conformation studies when only a sequence is known and sequence homologies in a sequence data bank could be rapidly searched for a biopolymer of known conformation.

There are two common styles for data representation in computer files. One style uses fixed length records with fields in defined positions. The advantages of using such a format are that it is easily adapted from existing formats, and programs which can read it are relatively easy to write. The inherent problems using defined position fields are that the addition of new fields means that all files in an older version of the format become unreadable, and editing errors involving column shifts are difficult for users to see and correct if many rows are involved.

The second style uses variable length records with labeled fields. The advantages of this style are that new fields are easy to add, errors cannot arise from shifted columns, and the file can be constructed with block structure spacing to aid the user in seeing file organization. The disadvantages using variable length records and labeled fields are that programs to read such a format are more difficult to write because they must use parsing techniques and more characters are required to store the same amount of information. The particular style employed depends on the relative emphasis given to human readability, computer readability and amenability to editing. With the speed and storage capacity of current computers it is feasible to sacrifice machine readability for human readability. It may also be useful for a standard to employ a style which would allow the file to be read by relational or object-oriented data base programs.

It is important that the standard, or standards, be enforceable by providing a verification mechanism. This would mean that integral to the adoption of a standard file format should be implementation of a program for verification that files conform to the standard format. Such a program would read files supposedly written in standard format and report compliance or noncompliance at three levels. The first level is verification to insure against simple editing mistakes. The second level is verification that the syntax of the format has been followed. The third level is verification that the description corresponds to an "acceptable" molecular model. The third level of verification will require that there be an agreement on the minimal content of a molecular description. The paradigm of such a verification program is a compiler. Indeed, if it is desirable for the standard to be adaptable and extensible, it should perhaps be viewed not simply as a file format but as a computer language for the description of molecules.

Literature Cited

1. Brown, I.D. *Acta Crystallogr.* **1983**, *A39*, 216–224.
2. Brown, I.D. *Acta Crystallogr.* **1988**, *A44*, 232.
3. George, D. G.; Mewes, H.W.; Kihara, H. *Protein Seq. Data Anal.* **1987**, *1*, 27–39.
4. Gund, P.; Barry, C.D.; Blaney, J.M.; Cohen, N.C. *J. Med. Chem.* **1988**, *31*, 2230–2234.

RECEIVED May 26, 1989

INDEXES

Author Index

Auneveux, Guillaume, 89
Barnard, John M., 76
Bebak, H., 105
Bruce, Everett A., 82
Burt, Katherine, 50
Buse, C., 105
Chabernaud, Françoise, 89
Cook, Anthony P. F., 50
Donner, W. T., 105
Garavelli, John S., 118
Gay, Jean-Pierre, 89
Higgins, Kevin M., 50
Hoever, P., 105
Hong, Richard S., 10
Hopkinson, Glen A., 50
Jacob, H., 105
Jochum, Clemens J., 76
Johnson, A. Peter, 50

Klaus, H., 105
Lawson, Alexander J., 41
Leilous, Craig, 82
Macko, John L., 59
Olszewski, Robert M., 82
Pesch, J., 105
Potenzone, Rudy, Jr., 82
Roemelt, J., 105
Schilling, P., 105
Seals, James V., 59
Singh, Gurmaj, 50
Smith, Dennis H., 18
Town, William G., 68
Warr, Wendy A., 1
Welford, Stephen M., 76
Woost, B., 105
Zirz, C., 105

Affiliation Index

Barnard Chemical Information Ltd., 76
BASF, 105
Bayer AG, 105
Beilstein Institut, 41,76
Chemical Abstracts Service, 59
Ciba Geigy, 105
ICI Pharmaceuticals, 1
Hampden Data Services, 68

Hawk Scientific Systems, 10
Molecular Design Ltd., 18
ORAC Ltd., 50
Polygen Corporation, 82
Sandoz, 105
Télésystèmes, 89
University of Illinois at Chicago, 118

Subject Index

A

Application software
 advantages, 22
 compatibility, 61
 disadvantages, 22
 evolution, 21
 standards, 25
 systems for information integration, 29
 vendor competition, 29
 wide variety, 29
Applications, nature of those existing, 12
Aromaticity, problems in computer
  representation, 33
Atom properties, problems in computer
  representation, 35

B

Beilstein Registry number, description, 41
Biopolymer sequence data, file format, 121
Bond properties, problems in computer
  representation, 35
Brookhaven protein data bank,
  description, 121
Browsing in computerized data bases,
  pitfalls, 42
Business needs in selling software, 29

C

Cambridge crystallographic data base,
  description, 120
Carbon-completeness of LN, 45
CAS ONLINE
 for searching the chemical literature, 4
 downloading and uploading structures, 4
CENTRUM
 as integration tool, 82
 overview, 83
Chemical Abstracts Service registry system, 2

Chemical diagrams, included in documents, 73
Chemical information, definition, 18
Chemical information manager, 85,87f
Chemical information processing tools
 CENTRUM, 82–88
 DARC, 89–104
 MOLKICK, 76–81
 SMD, 105–117
Chemical information systems, requirements,
  89,105
Chemical publishing environment, CENTRUM,
  85,86f
Chemical query editor programs, 77
Chemical representation, problems, 33
Chemical structure-DBMS data cross
  registration, 100f
Chemical structure browsing, 41–49
Chemical structure data
 integration with property data, 3
 standard format for exchange, 10
Chemical structure diagram, importance in
  chemistry, 69
Chemical structure drawing packages, 3
Chemical structure information
 levels of representation, 31
 preferred method of communicating, 1
Chemical structure queries, transmission, 78
Chemical structure searching, front-end
  programs, 77
Chemical structures, integration with
  text, 74
Chemical substances, classes for computer
  representation, 36
Chemists
 computing requirements, 69
 data bases, 71–72
Chemist's workstation, design
  requirements, 70
Classes of chemical substances, for computer
  representation, 36
Collaborative software developments, 5
Combined functionality, of LN, 45
Communication programs, PC-mainframe, 76

Communication with IMB PCs and
  Macintoshes, DARC, 101
Company data bases, for use by chemists, 72
Compatibility
  front-end software with online systems, 60
  networks, 61
  standard interface software, 61
Computational software, information to be
  maintained, 12
Computer Graphics Metafile (CGM) format,
  attempt at standardization, 16
Computer hardware, price-to-performance
  ratio, 18
Computer industry, general trends, 19
Computer manufacturers, competition, 23
Computer representation of structures,
  problems, 33
Computing power, growth, 68
Computing requirements, chemists, 69
Concordance numbers, description, 42
Connection tables
  advantages over other representations, 73
  compatibility, 62
Connectivity, definition, 69
Conversion programs, approach to transfer of
  molecular structure data, 118–119
Cross-registration of chemical structures
  and DBMS data, DARC, 99
Cross search
  between DARC and DBMS application, 99
  of chemical structures and DBMS data,
    DARC, 96
CROSSBOW, 2

D

DARC (description, acquisition, retrieval,
  and correlation system)
  communication with IMB PCs and
    Macintoshes, 101
  for searching the chemical literature, 4
DARC CHEMLINK, for offline queries, 5
DARC communication modules
  development criteria, 90
  display or print of DBMS data, 92
  functionality, 91
  operating system, 102
DARC–ORAC conversion, 53
Data and structure output using HLI, 55
Data base management system, search
  application, 97f
Data base searching using HLI, 54

Data base software, information to be
  maintained, 12
Data bases, for use by chemists, 71–72
Data record in SMD files, 108
Data representation, two common styles, 123
Data transfer, between software systems, 52
De facto standard environments, 68
Description, acquisition, retrieval, and
  correlation system, See DARC
Design considerations for multipurpose
  structure files, 10–17
DISPLAY CN module, DARC, 96,98f
DISPLAY module, DARC, 94
Display of chemical structures, DARC in
  DBMS, 94
Display of DBMS data, DARC, 93f
Display or print of DBMS data, DARC
  communication modules, 92
Distributed application computing
  environment, 84f
Document production, with PostScript, 26
Downloading, CAS ONLINE structures, 4
Drawing chemical structures, on graphics
  terminals and PCs, 4
Dyson notation, 2

E

Electronic transmission, chemical
  reports, 74
End-user, defining, 11
Enjoyability, as component of standard
  interface software, 64
Esthetics, as component of standard
  interface software, 64
External files, primary method for exchange
  of chemical structures, 32

F

File format, standard molecular description,
  119–121
Formula number, description, 42
Front-end software
  description, 4–5
  effect of changes to online systems, 60
  for chemical structure searching, 77
Front ends
  to CAS online, STN Express, 5
  to online systems, development, 3
Functionality, DARC communication
  modules, 91

G

Graphical objects, and internal
   representations, 32
Graphics emulation, terminal emulation
   programs, 77

H

Hardware
   DARC communication modules, 102
   diversity and incompatibility, 28
   multiplicity of vendors, 19
   revolution taking place, 19
   standards, 23
Hardware environments, compatibility, 61
Hardware revolution
   advantages, 19
   disadvantages, 20
Host Language Interface (HLI)
   description, 52
   potential applications, 57
   uses, 53–57

I

Identification number, description, 41
Information blocks
   for storage of molecular and reaction
     data, 108
   SMD structures, 107
Information exchange, between or among
   systems, 30
Information hierarchy, SMD structure,
   111,115
Information integration
   areas of progress, 38
   forces working against, 37
   gaps remaining, 29
   in an incompatible world, 18–40
   structures and data in different software
     systems, 30
   vendor lack of cooperation, 23
Integrated chemical and biological
   information system, components, 50
Integrated program system, scheme, 115f
Integration
   and standards, use of a host language
     interface, 50–58
   at search and display level, 52
   barriers, 29
   chemical structural information, 32

Integration—*Continued*
   chemical structures with text, 73
   definition, 69
   of various systems, abilities required, 51
   with synthesis planning systems, 55
Integration tools, SMD, 206
Interfacing, early attempts, 1
Intermediate standard format, to mitigate
   compatibility problems, 27
International support, of standard interface
   software, 63

L

Laboratory data bases, for use by
   chemists, 71
Lawson number
   carbon-completeness, 45
   combined functionality, 45
   description, 41
   design, 43
   example of use, 46
   factors governing its value, 44
   idea behind, 42
   limitations, 44
   limiting aspects, 45
   range searching, 48
   transparency, 45
   use like Beilstein Handbook, 43
   use of several in combination, 48
Line notations, for chemical structure
   information, 1
Litigation, by vendors protecting their
   technology, 24
Local data bases, for use by chemists, 71
Local edits, as component of standard
   interface software, 64

M

MACCS
   for chemical structure searching, 2
   for in-house chemical data bases, 4
Mass storage, information on disk or
   tape, 31
Memory, moving data into and out of, 32
Microcomputer, impact on chemical
   information, 3
Modularity, as component of standard
   interface software, 64
Molecular description file, development
   considerations, 122

Molecular description file formats, description, 120
Molecular graphics, evolution, 2
MOLKICK
chemical information processing tool, 76–81
query-drawing functions, 77
query editor, 5

N

National Institutes of Health/Environmental Protection Agency Chemical Information System, 2
Networks, compatibility, 61

O

Office document architecture, standard, 26
Offline query formulation, 4
Online access, to Beilstein files, 76
Online systems, effect of changes on front-end software, 60
Open architecture, definition, 69
Operating environments
advantages, 21
definition, 20
disadvantages, 21
standards emerging, 24
Operating systems
DARC communication modules, 102
five standards, 28
standards, lack of compatibility, 23
ORAC (Organic Reactions Accessed by Computer)
integration with other systems, 50–58
operating systems, lack of compatibility, 23
reaction indexing software, 4
OSAC (Organic Structures Accessed by Computer)
for chemical structure searching, 3
integration with other systems, 50–58

P

Personal computers
capabilities, 69
growth, 68
Pictorial representation
developing a standard, 13
of structures, storage, 12

PostScript, standard, 26
Printing of DBMS data, DARC, 95f
PSIDOM, PC-based package, 5
PsiGen, for structure drawing, 4
Public data bases, for use by chemists, 72

Q

Quasi-compatibility, difficulty in standard interface, 65
Query editor programs, 77

R

Range searching with the LN, 48
Reaction data bases, information to be maintained, 12
Real time and parallel process for display of DBMS data, DARC, 93f
Records, fixed vs. variable length, 123
Research computing environments, aspects, 69
Retrieval terms, in computerized data bases, 41
ROSDAL string format, 78–80

S

S4 search system, 78
SANDRA, structure and reference analyzer, 4
Screen format, definition using DARC, 96f
Search tool, LN, 43
Simplified molecular input line entry system (SMILES) notation, 2
SMD, See Standard molecular data
SMILES (simplified molecular input line entry system) notation, 2
Software
costs, ratio to hardware costs, 18
DARC communication modules, 102
for interfacing PC to major vendor, 4
vs. hardware advances, 19
Standard crystallographic file structure, 121
Standard environments, de facto, 68
Standard file structures, 6
Standard format
exchange of chemical structure data, 10
intermediate, to mitigate compatibility problems, 27
proposals, 122

Standard interface
  description, 59
  special difficulties, 65
  to public corporate and personal files, 59–67
Standard interface software
  international support, 63
  necessary features, 61–65
Standard molecular data (SMD)
  as standard for exchange of chemical structures, 60
  file as structural tools, 107
  file format, for interchange of chemical information, 121
  format, as integration tool, 105–117
  general strategy and fundamental properties, 106
  library of subprograms, 115
Standard molecular data (SMD) structure
  2-hydroxypyridine N-oxide, 109–110
  divisions, 107
  for a reaction, 113–114
  polyphenyl sulfide polymer, 111–112
Standard molecular description file format, to transfer molecular structure data, 119–120
Standardization
  early attempts, 1
  in the Macintosh environment, 6
Standards
  and vendor objections, 14
  as a way of reducing incompatibilities, 23
  competing, 16
  for pictorial representation, difficulties in developing, 13
  necessity, 59,118
  overly comprehensive, 16
  preference for choice, 15
  representation of chemical structures, 30
Stereochemistry, problems in computer representation, 33
STN Express, front end to CAS ONLINE, 5
Storage of pictorial representations of structures, 12
Structure and data output using HLI, 55
Structure and substructure search, among systems, 30
Structure browsing, and LN, 42
Structured query language standard, 25
Structures, different methods of representing, 30
Structure/substructure retrieval, 79f
Subblocks and superatoms, SMD structures, 107

Synthesis planning systems, integration with HLI, 55
System number, description, 41

T

Tagged image file format, standard, 26
Tautomerism, problems in computer representation, 34
Terminal emulation programs, 76–77
TOPFRAG, topological fragment code generator, 4
Topological fragment code generator, 4
Transferring molecular structure data, various approaches, 118
Transmission
  chemical structure queries, 78
  speed and noise, 66
Transmission of data, one way vs. two way, 17
Transparency, of LN, 45

U

Universal standards, 26
Universal structure/substructure representation, PC-host communication, 76–81
UNIX wars, 24
Uploading, CAS ONLINE structures, 4
User compatibility, importance, 62

V

Vendors
  approaches to hardware competition, 23
  competition in application software, 29
  hardware, multiplicity, 19
  lack of cooperation for information integration, 23
  litigation to protect their technology, 24
  objections to standards, 14

W

WIMPS (windowed, icon-based, mouse-driven, pointing systems) operating environment, 21
Wiswesser line notation, computerized, 2
Workstation, definition, 69

Production and Indexing: Janet S. Dodd

Acquisitions Editor: Cheryl Shanks

Elements typeset by Hot Type Ltd., Washington, DC
Printed and bound by Maple Press, York, PA

The chemical structure diagram is the chemist's preferred method of communication. With the advent of computer graphics, the proliferation of PCs, and the development of user-friendly software, the average chemist can now build chemical data bases, search structures and substructures, and generate chemical reports. However, this multitude of systems and data bases cannot be linked in a "seamless" manner because the software is being developed far ahead of the standards for the storage and exchange of data.

Time taken to input and manipulate structure data in several environments is not time well spent, nor is the time to learn how to use the different software packages required for structure drawing, online searching, data base management, and word processing. Ideally, neither commercial pressures nor technical complications should prevent a user from drawing a chemical structure the way he or she wants to and from accessing any personal, corporate, or public file. However, in practice, both commercial and technical factors cause severe limitations. A need exists for "seamless" interfaces and good communication among existing systems, but these goals cannot be accomplished until standards are established.

This book addresses these considerations in 12 chapters written by people from diverse backgrounds, from software developers to information specialists. Some of the specific topics covered are the need for flexibility in file formats to enable